MASSACHUSETTS INSTITUTE OF TECHNOLOGY

# AP LLO

## GUIDANCE, NAVIGATION
## AND CONTROL

E-2500

DESIGN SURVEY OF THE

APOLLO INERTIAL SUBSYSTEM

Edited by
William A. Stameris

MARCH 1970

# MIT    CHARLES STARK DRAPER
# LABORATORY

CAMBRIDGE, MASSACHUSETTS, 02139

ACKNOWLEDGEMENT

This report was compiled and edited by William A. Stameris and Paul F. Jopling of the Charles Stark Draper Laboratory, Massachusetts Institute of Technology. It was prepared under DSR Projects 55-31650 and 55-23870, sponsored by the Electronics Research Center and the Manned Spacecraft Center of the National Aeronautics and Space Administration under contracts NAS 12-642 and NAS 9-4065 with the Massachusetts Institute of Technology.

The publication of this report does not constitute approval by the National Aeronautics and Space Administration of the findings or the conclusions contained herein. It is published solely for the exchange and stimulation of ideas.

E - 2500

# INERTIAL SUBSYSTEM DESIGN SURVEY

## ABSTRACT

This document E - 2500 is a recorded history of the design and development of the Apollo inertial subsystem (CSM, LM) intended for manned spaceflight beyond earth orbits.

Its purpose is a source document for subsequent design criteria monographs.

Edited by
William A. Stameris
March 1970

# TABLE OF CONTENTS

## LIST OF ILLUSTRATIONS

# LIST OF ILLUSTRATIONS (CONT.)

ISS DESIGN SURVEY

Section I

INTRODUCTION

The Design Survey of the Apollo Inertial Subsystem was funded by NASA under Contract NAS-12-642. This contract was administered by the NASA Electronics Research Center of Cambridge, Massachusetts. The purpose of this document is to record the history of the Apollo Inertial Subsystem design and development (CSM and LM) in order to provide a source document for subsequent design criteria monographs.

The record provided by this document covers the design, development, and testing of an inertial subsystem intended for manned spaceflight beyond earth orbits. In addition, it highlights the experience and knowledge accumulated by this portion of the Apollo program. Finally, technical and program-oriented recommendations have evolved naturally from the experience documented and such recommendations are given as an important part of this Design Survey.

The text is structured to present certain sections with a narrative treatment of the entire topic. Section II, the Apollo Mission, introduces the reader to the overall mission plan within which the Apollo Inertial Subsystem is utilized. A description of how the inertial subsystem requirements evolved out of the overall mission plan follows. The reader who is familiar with Apollo may wish to omit the detailed mission description.

Section III, The Apollo Inertial Subsystem, an Overview, comprises one section wherein the present design and its evolution, a description of Inertial Subsystem Problems and Solutions, Assembly and Test Highlights, the Inertial Subsystem Reliability Assessment, and finally, Checkout and Flight Experience are presented. This section provides enough technical detail to be meaningful yet is general enough to be comprehended without requiring a specific background in the Apollo Program.

Section IV, Inertial Subsystem (ISS) Components, presents a more detailed treatment of individual components and can be used as a reference section to follow up on problems discussed in Section III.

Section V, System Flight Experience, reviews the salient features of the early manned flights.

Section VI, Conclusions, summarizes all of the material of the ISS Design Survey with regard to the lessons learned and implications for the future.

Section II

# THE APOLLO MISSION

## 2.1 Apollo Configuration

The goal of the Apollo Project is to place human exploration teams onto the moon and return them safely to earth. A spaceship consisting of three modules is launched on a trajectory to the moon by a Saturn V launch vehicle. The Command Module (CM) is designed for atmospheric re-entry and is to be the home for the three-man crew during most of the trip. The Service Module (SM) provides maneuver propulsion, power and expendable supplies. The Lunar Module (LM) is the vehicle which actually makes the lunar descent. It carries two of the three-man crew to the lunar surface while the other two modules remain in lunar orbit. The Apollo Guidance and Navigation System is the primary onboard equipment used for determination of the position and velocity of the lunar module and for control of its maneuvers. Similar guidance equipment is contained in both the Command Module and the Lunar Module. Each vehicle is equipped with a device for remembering spatial orientation and measuring acceleration, an optical angle device for angle measurements, displays and controls, means to interface with a spacecraft control system and indicators, and a central digital data processor.

To facilitate a greater understanding of the part played by the Apollo Inertial Subsystem, a description of the Apollo mission follows.

## 2.2 Mission Phases

The overall Apollo mission trajectory is summarized in Figure 2-1. The heavy lines in the figure correspond to the short thrusting maneuvers which are separated by the much longer free-coasting phases. The trajectory shown in Figure 2-1 is purposefully distorted to show the features of each phase more clearly.

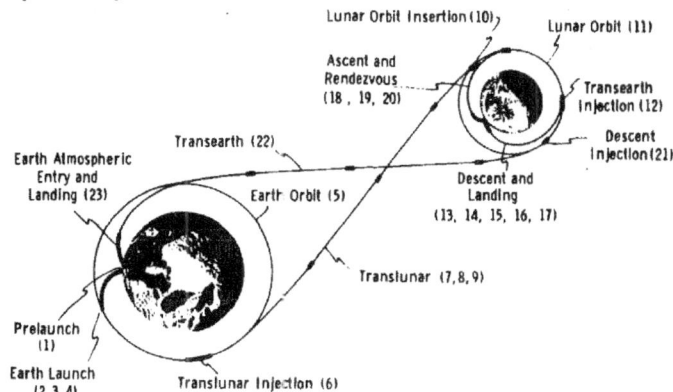

Fig. 2-1 The Overall Apollo Mission

The numbers relate to the mission-phase subdivisions which are discussed in detail in the following paragraphs.

2.2.1   Phase 1 -- Prelaunch

The prelaunch phase includes a schedule of activity to prepare and verify all equipment for flight. Automatic programmed checkout equipment is utilized to perform exhaustive tests of the major subassemblies.

Ground support equipment communicates directly with the Saturn and Apollo CM guidance computers to read in initial conditions and mission and trajectory constants as these parameters vary as a function of countdown status. Both sets of inertial guidance sensors are aligned to a common vertical and launch azimuth framework. The vertical is achieved in both cases by erection loops sensing gravity. Azimuth in Saturn is measured optically from the ground and controlled by means of an adjustable prism mounted on the stable member. Azimuth in Apollo is verified optically onboard by the astronauts and held by gyrocompassing action. During countdown, both systems are tied to an earth frame reference. Just before liftoff, both systems respond to signals to release the coordinate frames simultaneously from the earth reference to the nonrotating inertial reference to be used during boost flight. A third set of guidance equipment located in the Lunar Module (LM) is used near to the moon.

2.2.2   Phase 2 -- Earth Launch, First Stage

During first stage flight, the Saturn guidance system controls the vehicle by swiveling the outer four rocket engines. During the initial vertical flight, the vehicle is rolled from its launch azimuth to the flight path azimuth. Following this maneuver, the Saturn guidance system controls the vehicle in an open-loop preprogrammed pitch maneuver which is designed to pass safely through the period of high aerodynamic loading. Inertially-sensed acceleration signals are not used during this phase to guide the vehicle to the desired flight path, but rather lateral accelerometers aid in controlling the vehicle to stay within the maximum allowable angle of attack. Stable control is achieved in overcoming the effects of flexure bending, fuel slosh, and aerodynamic loading by the use of properly located sensors and control networks.

Both the Saturn and Apollo Command Module guidance systems continuously measure vehicle motion and compute position and velocity. In addition, the Apollo system compares the actual motion history with that to be expected from the Saturn control equations and generates an error display to the crew. This and many other sensing and display arrangements monitor the flight. If abort criteria are exceeded, the crew can fire the launch escape system. This escape system consists of a rocket on a tower attached to the top of the Command Module, which has the capability of lifting the crew rapidly away from the rest of the vehicle. Parachutes would be later deployed for landing.

In a normal flight, the first stage is allowed to burn to nearly complete fuel depletion as sensed by fuel level meters before first stage engine shutdown is commanded.

### 2.2.3 Phase 3 -- Earth Launch, Second Stage

Shortly after the initial fuel settling ullage and the firing of second stage thrust, the aerodynamic pressure decreases to zero as the vehicle passes out of the atmosphere. At this time the launch escape system is jettisoned. Aborts at this time, if necessary, would normally be accomplished using the Apollo Service Module propulsion to accelerate the Command Module away from the rest of the vehicle.

Since the problems of aerodynamic structure loading are unimportant in second stage flight, the Saturn guidance system steers the vehicle towards the desired orbital insertion conditions using propellant-optimizing guidance equations. Thrust control is achieved by swiveling the outer four engines of the second stage.

During second-stage flight, the Apollo Command Module guidance system continues to compute vehicle position and velocity. Also, this system computes any of several other possible parameters of the flight to be displayed to the crew for monitoring purposes. In addition, the free-fall time to atmospheric entry and the corresponding entry-peak acceleration are displayed to allow the crew to judge the abort conditions existing.

### 2.2.4 Phase 4 -- Earth Launch, Third Stage

The third Saturn stage or SIVB has a single engine for main propulsion which is gimballed for thrust-vector control. Roll control is achieved by use of the SIVB roll-attitude control thrusters.

The Saturn guidance system continues to steer the vehicle to orbital altitude and speed. When orbit is achieved, the main SIVB propulsion is shut down.

During second and third stage boost flight, the Apollo Command Module has the capability, on astronaut option, to take over the SIVB stage guidance function if the Saturn guidance system indicates failure. If this switchover occurs, the mission presumably could be continued. More drastic failures would require an abort using the Service Module propulsion.  In such a case, the Apollo guidance computer would be programmed to provide a number of abort trajectories:  (1) immediate safe return to earth, (2) return to a designated landing site, or (3) orbit for later return to earth.

SIVB engine shutdown occurs about 12 minutes after liftoff at 100 n. mi. altitude near circular orbit.

### 2.2.5   Phase 5 -- Earth Orbit
The Apollo spacecraft configuration remains attached to the Saturn SIVB stage in earth orbit.  The Saturn system controls attitude by on-off commands to two of the small fixed attitude thrusters for pitch and to four more shared for yaw and roll.

Ground tracking navigation data telemetered from the Manned Space Flight Network (MSFN) stations are available to correct the position and velocity of the Saturn navigation system and provide navigation data for the Apollo navigation system.  In Apollo, the crew can also make navigation measurements for onboard determination of the ephemeris by making landmark or horizon direction sightings using a special optical system.  The Apollo inertial equipment alignment is updated by star sightings with the same optical system.  For these measurements the crew has manual command-control of attitude through the Saturn system.  Normally, limited roll maneuvers are required to provide optical system visibility to both stars and earth.

Typically, the earth-orbital phase lasts for several hours before the crew signals the Saturn system to initiate the translunar injection.

### 2.2.6   Phase 6 -- Translunar Injection
Translunar injection is performed using a second burn of the Saturn SIVB propulsion, preceded by an ullage maneuver using the small thrusters.   Saturn guidance and control systems again provide the necessary steering and thrust vector control to the near parabolic velocity which, for crew safety considerations, put the vehicle on a so-called "free return" trajectory to the moon.  The system aims to this trajectory which ideally is constrained to pass in back of the moon and return to earth-entry conditions without additional propulsion.

As before, the Apollo guidance system independently generates appropriate parameters for display to the crew for monitoring purposes.  If the Saturn guidance system indicates failure, steering takeover by the Apollo crew is possible without need for aborting the mission.

## 2.2.7 Phase 7 -- Transposition and Docking

The spacecraft configuration injected onto the translunar free-fall path must be reassembled for the remaining operations.

The astronaut pilot separates the Command and Service Modules (CSM) from the SIVB. He then turns the CSM around for docking to the LM which is housed inside the adapter in front of the SIVB stage. To accomplish this, the pilot has a three-axis, left-hand translational controller and a three-axis, right-hand rotational controller. For the maneuver, output signals from these controllers are processed to appropriately modulate the firing of the sixteen low-thrust reaction control jets of the Service Module. The normal response from the translational controller is proportional vehicle acceleration in the indicated direction. The normal response from the rotational controller is proportional vehicle angular velocity about the indicated axis.

During the separation and turnaround maneuver of the CSM, the SIVB control system holds the LM attitude stationary to permit a simple docking maneuver of the command module to the LM docking hatch. The SIVB, Saturn instrument unit and LM adapter are staged to leave the Apollo spacecraft in the translunar configuration. Final docking is completed less than 6.5 hours from liftoff.

## 2.2.8 Phase 8 -- Translunar Coast

Soon after injection into the translunar free-fall coast phase, navigation measurements are made and processed to examine the acceptability of the trajectory. If these data indicate the need, an early midcourse maneuver corrects the flight path error before it propagates with time into larger values which would needlessly waste correction-maneuver fuel.

Once this first correction is made -- perhaps a couple of hours from injection -- the navigation activity on board proceeds at a more leisurely pace. Ground tracking data can be telemetered to the craft when available. Using these ground data and/or onboard sextant-type landmark-to-star angle measurements, the onboard computer corrects the spacecraft state vector position and velocity information.

The astronaut navigator can examine, with the aid of the computer, each datum input available -- whether from ground tracking telemetered to the craft or taken onboard -- to see how it could change the indicated position and velocity before it is accepted into the computer state vector correction program. In this way, the effects of mistakes in data gathering or transmission can be minimized.

2.2.9  Phase 9 -- Midcourse Corrections

The navigator periodically examines the computer estimate of indicated uncertainty in position and velocity and the estimate of indicated velocity correction required to improve the present trajectory. If the indicated position and velocity uncertainty is suitably small and the indicated correction is large enough to be worth the effort in making, then the crew prepares for the indicated midcourse correction. Each midcourse velocity correction requires initial spacecraft orientation to put the estimated direction of the thrust axis along the desired acceleration direction. Once thrust direction is attained, the propulsion system is fired under the measurement and control of the guidance system. Use of the guidance system requires that the inertial measurement system be aligned by optical star-direction sightings.

Typical midcourse corrections are expected to be of the order of 30 feet per second. If the required correction happens to be very small, it is made by using the small reaction control thrusters. Larger corrections are made with a short burn of the main service propulsion engine. About three such midcourse velocity corrections are required on a trip to the moon. The direction and magnitude of each correction adjust the trajectory so that the moon is finally approached near a desired plane and pericynthian altitude which provides for satisfactory conditions for the lunar orbit insertion.

2.2.10  Phase 10 -- Lunar Orbit Insertion

For lunar orbit insertion, as in all normal thrusting maneuvers with the service propulsion engine, the inertial guidance system is first aligned using star sightings. Then the system generates initial conditions and steering parameters based upon the navigation measure of position and velocity and the requirements of the maneuver. The guidance initiates engine turn-on, appropriately controls the direction of the acceleration, and signals engine shut-down when the maneuver is complete.

The lunar orbit insertion maneuver is intended to put the spacecraft in a near-circular orbit of approximately 80 n. mi. altitude. The plane of the orbit is selected to pass over the landing region on the front of the moon.

2.2.11  Phase 11 -- Lunar Orbit

In lunar orbit, navigation measurements are made to update the actual orbital motion information. The navigation measurement data are processed in the computer using much of the same program used in the translunar phase. Several sources of data are possible. Direction measurement to lunar landmarks or horizons and earth-based radio tracking telemetered data are similar to the measurements used earlier in the flight in earth orbit. Because of the lack of lunar atmosphere, the times of selected star occultations by the lunar limb can conveniently be measured. Orbital period measurements are available by timing successive passages over the same terrain feature or successive occultations of the same star. Enough measurements must be made to provide accurate initial conditions for the guidance system in the LM for its controlled descent to the lunar surface. Before separation of the LM, the landing area is examined by the crew using the magnifying optics in the command module. At this time, direction measurements to a particular surface feature can relate a desired landing site or area to the existing indicated orbital ephemeris in the computer. These particular landing coordinates become part of the LM guidance system initial conditions received from the command module.

After two of the crew transfer to the LM and separate from the Command and Service Module (CSM), the remaining man in the CSM continues orbital navigation as necessary to maintain an accurate knowledge of the CSM position and velocity.

2.2.12 Phase 12 -- LM Descent Orbital Insertion

The LM guidance system is turned on and checked out earlier in lunar orbit before separation and before initial conditions are received from the CSM. About twenty minutes before initiation of the LM descent injection maneuver, the vehicles are separated, the LM guidance system receives final alignment from star sightings, and the attitude for the maneuver is assumed. The maneuver is made using the LM descent stage propulsion under control of the LM guidance system. The throttling capability of the descent engine is exercised as a check of its operation during the short burn. The maneuver is a 100 ft. per second velocity change to reduce the 5250 ft. per second orbital velocity for a near-Hohmann transfer to 8 n. mi. altitude pericynthian which is timed to occur at a range of about 200 n. mi. short of the final landing area.

2.2.13 Phase 13 -- LM Descent Orbit Coast

During the free-fall phases of the LM descent, the CSM can make tracking measurements of the LM direction to confirm the LM orbit with respect to the CM. For that part of the trajectory in the front of the moon, earth tracking provides an independent check. The LM, during appropriate intervals in the coasting orbit, checks the operation of its radar equipment. The directional tracking and ranging operation of the rendezvous radar is checked against the radar transponder on the CSM. These data provide the LM computer with added information for a descent orbit check. At lower altitudes, the LM landing radar on the descent stage is operated for checks using the moon surface return. Alignment updating of the LM guidance system is performed if desired. The CM from its orbit, using the tracking systems and onboard computer, can monitor this phase of the LM descent. As pericynthian is approached, the correct LM attitude for the powered descent phase is achieved by signals from the guidance system.

## 2.2.14  Phase 14 -- LM Powered Descent Braking Phase

This phase starts at the 8 n. mi. altitude pericynthian of the descent coast phase. The descent engine is reignited, and the velocity and altitude-reducing maneuver is controlled by the LM inertial guidance and control system.

The descent stage engine is capable of thrust level throttling over the range necessary to provide initial braking and controlled hover above the lunar surface. Engine throttle setting is commanded by the guidance and control system to achieve proper path control, although the pilot can override this signal if desired.

Thrust vector direction control of the descent stage is achieved by a combination of body-fixed reaction jets and limited gimballing of the engine. The engine gimbal angles follow guidance commands in a slow loop so as to cause the thrust direction to pass through the vehicle center of gravity, thus minimizing the need for continuous fuel-wasting torques from the reaction jets. During all phases of the descent, the operations of the various systems are monitored. The mission could be aborted for a number of reasons. If the primary guidance and control system performing the descent control is still operating satisfactorily, it would control the abort back to rendezvous with the CSM. If the primary system has failed, a simple independent abort guidance system can steer the vehicle back to conditions for rendezvous. For a normal mission, however, the braking phase continues until the altitude drops to about 4 n. mi. Then guidance control and trajectory enter the final approach operation.

## 2.2.15  Phase 15 -- LM Powered Descent Final Approach

One significant feature of this phase is that the controlled trajectory is selected to provide visibility of the landing area to the LM crew. The vehicle attitude, descent rate, and direction of flight are controlled by the guidance so that the landing site appears fixed with relation to the window. A simple reticle pattern in the window gives an indication of whether the landing point is aligned with the number indicated by the computer display. The pilot may observe that the landing point indicated, with relation to other areas nearby, is in an area of unsatisfactory features. The pilot can then elect to select a new landing point for the computer control by turning the vehicle about the thrust axis until the reticle intersects the better area. He then hits a "mark" button to signal the computer, reads the reticle

2-8

number which is in line with this area into the computer and then the guidance system appropriately redirects the path. This capability allows early change of landing area and efficient fuel control during acquisition of the new area which otherwise might have to be performed wastefully later during hover.

Automatic guidance control during the terminal phase uses weighted combinations of inertial sensing and landing radar data. The weighting depends upon expected uncertainties in the measurements. The landing radar includes altitude measurement and a three-beam Doppler measurement of three components of LM velocity with respect to the lunar surface.

2.2.16 Phase 16 -- Landing and Touchdown

At any point in the landing, the pilot can elect to take over partial or complete control of the vehicle. For instance, one logical mixed mode would have altitude descent rate controlled automatically by modulation of the thrust magnitude and pilot manual control of attitude for maneuvering horizontally.

The final approach phase ends near the lunar surface, and the spacecraft enters a hover phase. This phase can have various possibilities of initial altitude and forward velocity depending upon mission ground rules, pilot option and observed local terrain features. Descent stage fuel allowance provides for approximately two minutes of hover before touchdown. This must be accomplished in the time allotted or abort on the ascent stage is initiated. The crew will make final selection of the landing point and maneuver to it either by tilting the vehicle or by operating the reaction jets for translation acceleration. Inasmuch as the flying dust and debris resulting from rocket exhaust degrades radar and visual information, the inertial system altitude and velocity computation is updated by the landing radar so that, as touchdown is approached, good data are available from the inertial sensors. Touchdown must be made with the craft near vertical and at sufficiently low velocity.

2.2.17 Phase 17 -- Lunar Surface Operations

Time spent on the moon includes considerable exploration, experimentation and gathering of soil samples. Also, during the stay time, the LM spacecraft systems are checked and prepared for the return voyage. The ephemeris of the CSM in orbit is continually updated and the information relayed to the LM crew and computer. The LM rendezvous radar also can track the CSM as it passes overhead to provide further data upon which to base the ascent guidance parameters. The inertial guidance system receives final alignment utilizing optical star-direction sightings prior to the start of ascent. The vertical components of this alignment could also be achieved by accelerometer sensing of lunar gravity in a vertical erection loop. Liftoff is timed to achieve the desired trajectory for rendezvous with the CSM.

### 2.2.18 Phase 18 -- LM Ascent

LM powered ascent is accomplished with the LM ascent engine. The ascent engine, which has a fixed mounted nozzle, is under control of the LM inertial guidance and control system. Thrust vector is achieved by operation of the sixteen reaction jets that are mounted on the ascent stage. The engine cannot be throttled, but the necessary signals from guidance will terminate the burning when a suitable trajectory is achieved.

The initial part of the ascent trajectory is a vertical rise followed by a pitchover as commanded by the guidance equations. During the pitchover maneuver, the guidance system also commands roll steering such that the LM maneuvers into an elliptical orbit coplanar with the CSM plane.

### 2.2.19 Phase 19 -- LM Midcourse Maneuvers

A constant delta height maneuver is initiated at the apocynthion of the LM concentric elliptical orbit that places the LM in a circular orbit which is predetermined such that the LM arrives at a required position with respect to the CSM at the terminal phase initiation time. During this phase, the crew verifies LM trajectory parameters as computed by the LM guidance computer with the parameters computed by the command module computer.

### 2.2.20 Phase 20 -- Terminal Rendezvous and Docking

The terminal rendezvous phase consists of a series of braking thrust maneuvers under control of the LM guidance system which uses data from its inertial sensors and the rendezvous radar. The objective of these operations is to reduce the velocity of the LM relative to the CSM to zero at a point near the CSM. This leaves the pilot in the LM in a position to initiate a manual docking maneuver with the CSM using the translation and rotation control of the LM reaction jets. Although these maneuvers would normally be done with the LM, propulsion or control problems in the LM could require the CSM to take the active role.

After final docking, the LM crew transfers to the CSM. The LM is then jettisoned and abandoned.

### 2.2.21 Phase 21 -- Transearth Injection

Navigation measurements made while in lunar orbit determine the proper initial conditions for transearth injection. These are performed as before using onboard and ground-based tracking data as available.

The guided transearth injection maneuver is made normally under the control of the primary inertial guidance system. Several backup means are available to cover possible failures in the primary system. The injection maneuver is controlled to put the spacecraft on a free-fall coast to satisfactory entry conditions near earth. The time of midcourse transearth coast must be adjusted by this injection maneuver to account for earth's rotation motion of the recovery area and as limited by the entry maneuver capability.

## 2.2.22 Phase 22 -- Transearth Coast

The transearth coast is very similar to the translunar coast phase. During the long coasting phases going to and from the moon, the systems and crew must control the spacecraft orientation as required. Typical midcourse orientation constraints are those necessary to assure that the high-gain communication antenna is within its gimbal limits to point to earth, or that the spacecraft attitude is not held fixed to the local heating effect of the sun for too long a period.

During the long periods of free-fall flight to and from the moon, when the inertial measurement system is not being used for controlling velocity corrections, the inertial system is turned off to conserve power supply energy.

Onboard and ground-based measurements provide for navigation upon which is based a series (normally three) of midcourse correction maneuvers during transearth flight. The aim point of these corrections is the center of the safe earth entry corridor suitable for the desired landing area. This safe corridor is expressed as a variation of approximately ±17 n. mi. in the vacuum perigee. A too-high entry might lead to an uncontrolled skip out of the atmosphere. A too-low entry might lead to atmospheric drag accelerations exceeding the tolerance of the crew.

After final safe entry conditions are confirmed by the navigation system prior to the entry phase, the inertial guidance system is aligned, the Service Module is jettisoned and the initial entry attitude of the Command Module is attained.

## 2.2.23 Phase 23 -- Earth Atmospheric Entry

Initial control of entry attitudes is achieved by guidance system commands to the twelve reaction jets on the command module surface. As the atmosphere is entered, aerodynamic forces create torques determined by the shape of the CM and center-of-mass location. If initial orientation was correct, these torques are in a direction towards a stable trim orientation with heat shield forward and flight path nearly parallel to one edge of the conical surface. The control system now operates the reaction jets to damp out oscillation about this trim orientation. The resulting angle of attack of the entry shape causes an aerodynamic lift force which can be used for entry path control by rolling the vehicle about the wind axis under control of the guidance system. Range control is achieved by rolling so that an appropriate component of that lift is either up or down as required. Track or across range control is achieved by alternately choosing as required the side on which the horizontal lift component appears.

The early part of the entry guidance is concerned with the safe reduction of the high velocity through the energy dissipation effect of drag forces. Later, at lower velocity, the objective of controlling the vehicle to the earth recovery landing area is included in the guidance programming. This control continues until the velocity is reduced and a suitable position is achieved for the deployment of a drogue parachute. Final letdown to a water landing is normally achieved by three parachutes.

Section III

# THE APOLLO INERTIAL SUBSYSTEM, AN OVERVIEW

## 3.1  Early Decisions --Block I

During 1957 through 1959, MIT/IL performed a study which led to the publi-
cation of a report in June of 1959 entitled "Recoverable Interplanetary Space Probe."
In September 1959, under a NASA contract, MIT/IL undertook a study of the
guidance and control design for a variety of space missions.  The study report was
issued in April 1960 and showed the onboard sensor capability which could support
Navigation and Guidance for manned or unmanned spacecraft.  During this period,
NASA was defining its advanced manned space flight program which would follow
Mercury and Gemini and which was named "Apollo" in July of 1960 by the NASA
Administrator.  The program was to lead, eventually, to a manned lunar landing.

In November 1960, a six-month contract with MIT to conduct preliminary
design work for the navigation and guidance support for project Apollo was
proposed by NASA.  The manned lunar landing was made a national goal by
President Kennedy in May 1961.  In August of 1961, NASA awarded its
first major Apollo contract to MIT/IL to develop the guidance and navigation system.
It is important at the outset to understand the factors governing the design of the
Apollo G&N in its initial phases.  In late 1961, plans were to fly an Apollo spacecraft in the fall of 1963 with an MIT G&N system on board.  The NASA direction
to support this schedule was to (a) build that which we knew how to build in the time
allotted, and (b) to use only components proven by production and operational
experience.

Numerous fundamental and far reaching decisions were made quickly in the
early days of the contract.  The IMU gimbal question was one of the most fundamental.  It was reasoned that IMU in-flight alignment would be a necessity.  No
gyro would be able to maintain adequate inertial reference for a period of twelve
days to cover all the necessary maneuvers.  With a realignment concept and
recognition that all maneuvers where the IMU was required were in-plane maneuvers
with little or no out-of-plane steering, it was reasoned that a 3 gimbal system could
be used.  This had several advantages over a 4 gimbal IMU in terms of system
complexity, weight, power, reliability and cost.

### 3.1.1 Gimbal System Arrangement

Although a strapdown or body-mounted inertial subsystem configuration was briefly examined at the start of the program, no serious consideration was given to this technique. The small development time that the schedule allowed and the fact that no such body-mounted system was yet out of the laboratory experimental stage precluded its use. Moreover, it was clear that schedules could be met by the existing development team by utilizing past experience with the design of the gimballed inertial measurement unit of the Polaris Mark 2 guidance system.

At the time the decision was made in 1961, there appeared to be no doubt that a three degree-of-freedom gimbal system would be the optimum choice for the inertial measurement unit. To simplify hardware design, the various ground rules mentioned earlier emphasized simplicity, light weight, optimum performance and utilization of crew capability. These ground rules all indicated that, for the Apollo mission, a three degree-of-freedom gimbal arrangement was to be preferred over the four degree-of-freedom configuration.

The function of the gimbal system arrangement was to support the gyros and accelerometers on a structure that would be kept nonrotating in space in spite of rotations of the spacecraft. The only motivation for a four degree-of-freedom gimbal system was that such a configuration could be made and operated so that all attitudes of the spacecraft would be accommodated without the problem associated with gimbal lock that could possibly occur with a three degree-of-freedom system. The question posed in 1961 and 1962 was: Could the simpler three degree-of-freedom IMU meet all the requirements Apollo would ask of it without danger of approaching gimbal lock and loss of stable member alignment? The answer in brief was that all usual Apollo maneuvers would naturally be such that gimbal lock would be avoidable by a sufficiently wide margin to be easily accommodated by the three degree-of-freedom candidate IMU under consideration. In unusual situations the operation near gimbal lock in nonemergency maneuvers could be simply avoided. Direct means were available to warn of approaching difficulty so that corrective action could be taken. And finally, the procedure for recovery from loss of alignment, if it occurred--even in emergency situations--seemed straightforward. The discussions that were relevant to IMU selection are included in the following sections.

### 3.1.2 Pertinent IMU Design Characteristics

The IMU design under consideration borrowed heavily from that of the Polaris Mark 2. The capability of this class of design to maintain stable member orientation in spite of base motion angles and rotational rates was well known. The Polaris IMU was a three degree-of-freedom gimballed structure such as shown in the "stick and wire" schematic form in Figure 3-1, and the proposed Apollo IMU was to be the same. Structural features of the gimbals that were developed for Apollo can be seen in Figure 3-2.

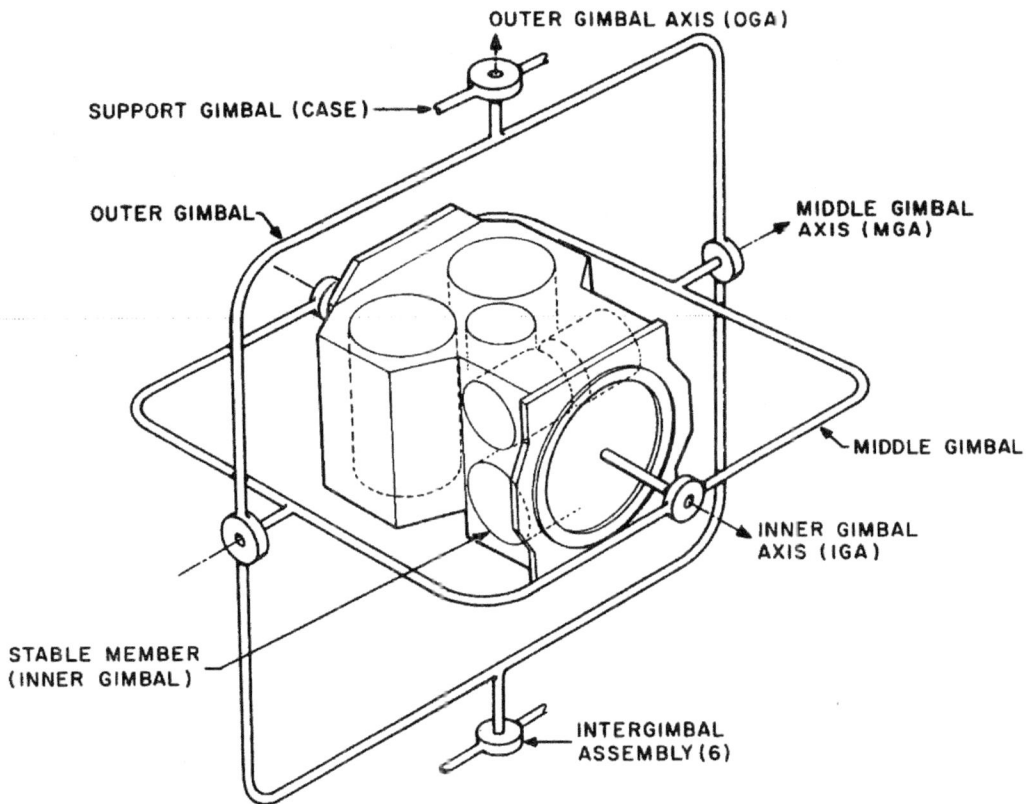

Fig. 3-1 Schematic of Inertial Measurement Unit

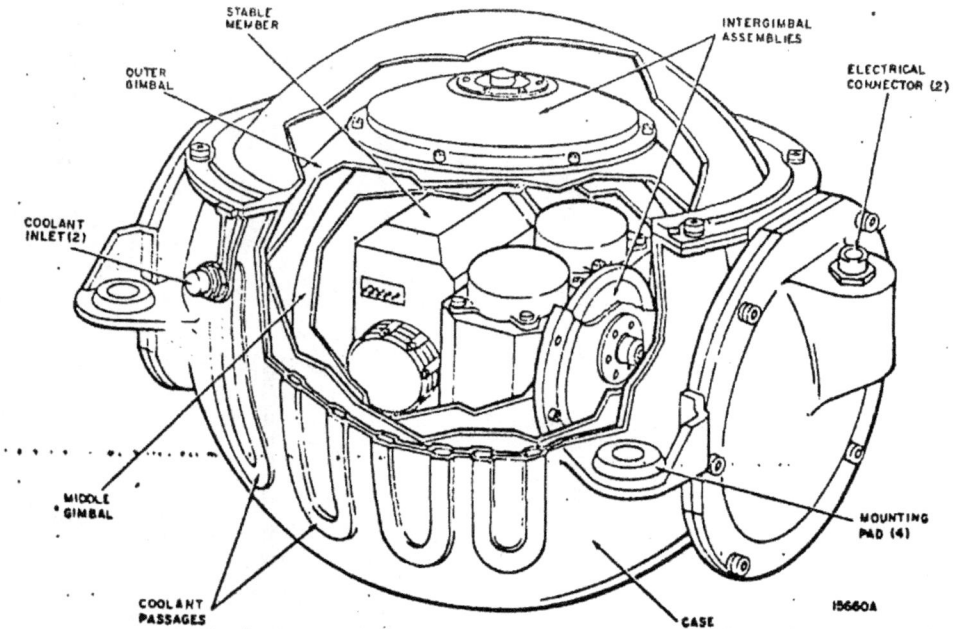

Fig. 3-2 Inertial Measurement Unit

Each gimbal axis of the proposed IMU had the servo torque motors and electromagnetic data transducers directly coupled to the two adjacent members without operating through a gear train such as is common in many gimbal designs. This operation without gears had a multifold advantage. Most obvious was the elimination of concern about gear wear and accuracy of mesh which were critical factors in making a reliable servo of the necessary high performance. Equally important was elimination of the requirement for the servo to provide torque to accelerate gear train inertia when no angular velocity was desired of the driven gimbal. Without gear trains the inertia of the stable member would tend to hold itself stationary without any help from the servo against any magnitude of base rotation motion except for the effects of (1) bearing friction, (2) motor "back emf" and (3) outer gimbal inertia. Discounting these last two effects for the moment, the servo would need to overcome only the small bearing friction of the Apollo IMU gimbals, even for extremely violent base or vehicle rotations. There are none of the limitations resulting from high angular rates or accelerations of the vehicle imposed by the usual gimbal axis gear trains.

The second effect mentioned above, back emf, concerns the voltage generator action of the motor, the servo amplifier drive impedance, a corresponding lag term in the servo loop, and a base motion coupling. The level of the output current feedback under consideration in the Apollo IMU servo would control these effects adequately so that any concern of base motion angular velocity coupling as a result of motor voltage would be trivial within any conceivable uncontrolled vehicle maneuvers from which recovery is possible.

The third effect, resulting from outer gimbal inertia, comes into play only with large middle gimbal angles away from the zero orientation. See Figure 3-1. In the extreme situation, large middle gimbal angles cause the effect called "gimbal" lock. This subject is so extensive that the ensuing section has been devoted to its exposition.

3.1.2.1 The Problem of Gimbal Lock. Gimbal lock occurs when the outer gimbal axis is carried around by vehicle motion to be parallel to the inner gimbal axis. At this point the three gimbal axes lie in a single plane. No gimbal freedom now exists to "unwind" base motion about an axis normal to this plane. Even though any vehicle orientation with respect to the stable member can be accommodated by particular sets of the three gimbal angles, the condition at gimbal lock prevents accommodation of a particular orientation change from the locked condition without exceedingly high outer gimbal acceleration.

Problems with a three degree-of-freedom system like the Apollo IMU which was being considered could occur under circumstances other than the gimbal locked situation. For instance, when the locked configuration is approached, as manifested when middle gimbal angles approach $90^{\circ}$, the stabilization capabilities of the assembly become more and more marginal depending on design. With proper gyro error signal resolution and gain control, the locked configuration can be very closely approached without undesirable effects. However, as gimbal lock is approached more closely, higher and higher angular accelerations are required of the outer gimbal to hold the inner member fixed against particular components of base angular velocity. By an inherent tendency to stay fixed, the inertias of the inner gimbals can generate much of the necessary reaction torque to help provide the required acceleration of the outer gimbal over a limited range. This would be perfect without the accelerating torque from the outer servo motor if either the inner structures were of infinite inertia or the outer structure were zero inertia.

Lacking an infinite ratio of the two, the remaining burden of providing the accelerating torque must be taken by the servo and outer gimbal motor.

In Apollo, the outer gimbal structure achieved necessary structural stiffness through the thin section spherical shape with a relatively small inertia. The inner stable member carried all the mass of the inertial components and necessary thermal sink mounting block.

Besides favorable inertia ratios, much of the capability of the Apollo IMU to handle near gimbal lock conditions was at first attributed to the use of a small angular accelerometer on each axis (ADA) as a servo stabilization feedback element. This device permits very high torque gains with low servo noise problems over all frequencies and allows specification operation over a wide margin of gain change. No critical adjustments are necessary. The Block I IMU utilized this ADA as the high frequency feedback device with good success. Later in the studies for the Block II IMU, the gyro noise inserted into the loop resulting from the necessary high frequency gain was found tolerable and the ADA was eliminated.

Moreover, it was argued that transient loss of attitude resulting from gimbal lock effects, or any other disturbance for that matter, did not necessarily mean a permanent loss of orientation indication of the IMU, unless the gyro gimbal output axis limit stops were reached. Within the integration range of the gyro, the attitude is recovered as the gyro error is brought back to zero. In effect, the orientation is memorized in the gyro output axis angle and angular velocity until the servo can recover.

Firm ground rules on how close to gimbal lock the Apollo IMU could operate satisfactorily depended upon experimental results with the actual flight configuration IMU. Data, using IMU #3 with breadboard electronics, were collected. From this information, it appeared that gimbal lock could be approached as closely as 10 degrees without risk and even much closer with some possibility of loss of stable member attitude. Stated more graphically, a test was conducted where the inner axis of the system was aligned within 10 degrees of a base motion axis perpendicular to the output axis. Base motion angular velocity then caused gimbal lock to be passed within 10 degrees. Stable member attitude was held consistently for this configuration with base angular velocities of 60 degrees per second. A tentative and conservative listing of acceptable vehicle rates and accelerations was then generated:

Vehicle Angular Velocity Allowable

| | |
|---|---|
| About inner gimbal axis (continuous) | 720 deg/sec[*] |
| About middle gimbal axis ($\pm 80^\circ$) | 720 deg/sec[*] |
| About outer gimbal axis (continuous) | 720 deg/sec[*] |
| About any arbitrary body or inertial axis resulting in passing up to $10^\circ$ of gimbal lock (continuous) | 60 deg/sec |

Vehicle Angular Acceleration Allowable

| | |
|---|---|
| About any axis and within above rate limits | $360 \text{ deg/sec}^2$[*] |

[*]Values marked with an asterisk have much higher limits, but were as yet undetermined.

3.1.2.2 <u>IMU Operation to Avoid Gimbal Lock</u>. Although the allowable vehicle motions described above to avoid effects near gimbal lock were less constricting than might be expected, the argument had to be made that the area near gimbal lock could be avoided in practical operation.

It was proposed that the Apollo IMU would normally be shut down during all long periods not requiring its use. This would be done primarily to save power and corresponding fuel cell battery reactant (estimated saving of 43 pounds of reactant in a 200 hour command module lunar landing mission). For this reason, and because of unavoidable angle drift over long time intervals resulting from imperfect performance of the gyros, the guidance system would provide for inflight IMU alignment against star references before the start of each accelerated phase of the mission. This allowed the IMU stable member alignment to be chosen for each use to the most logical orientation.

The decision to utilize only three degrees of IMU gimbal freedom did not elicit any counter arguments within NASA or from other contractors. The configuration was, for the most part, frozen long before the techniques called the "concentric flight plan" were developed from the rendezvous experience of Gemini where maneuvers were designed not for efficiency but for conceptual simplicity and ease of monitoring. Correspondingly, later planning, without a view to greater efficiency, identified pure plane changing maneuvers with an in-plane burn component.

Thus, the possibility of exactly out-of-plane burns became far more probable than was thought earlier to be the case.

Currently, the out-of-plane burns during rendezvous are accomplished with the RCS thrusters in a low acceleration translation. In this way, the vehicle can be accelerated sideways without changing spacecraft orientation to gimbal lock attitudes. This method also allowed for a maneuver which did not break radar lock or astronaut line of sight on the CSM. Avoidance of any gimbal lock problem during the actual thrusting and entry phases of Apollo offers no difficulty.

For the majority of mission aborts, the IMU gimbal lock situation was seen to put no more constraint on successful abort than on normal mission phases. Two mission phases were identified, however, which would require critical dependence upon the IMU indicated attitude to cope successfully with the emergency. The first of these was high altitude abort prior to launch escape tower jettison. If the command module tumbles during this operation there is a possibility that the outer axis might pass through the critical areas near gimbal lock causing the loss of IMU attitude information. If correct entry attitude is not assumed early enough, the reaction jets might not be able to overcome and correct the improper but stable attitude of having the conical end point forward as contrasted with the proper blunt end forward entry orientation. In view of the rate limitations of the backup attitude system and the problems of getting attitude cues through the windows, dependency on IMU attitude information might be necessary. The probability of exceeding crew stress limits by a conical end forward entry in the situation described is the product of: (1) the probability that abort is initiated at the critical altitude, (2) the probability that the abort initiates an uncontrolled tumble, (3) the probability that the tumble causes the IMU to pass through its critical gimbal lock areas, and (4) the probability that the pilot cannot sense the direction of drag early enough to correct attitude or that the command module enters conical end first. It was conjectured but not proven that the resulting probability was an extremely small number.

The second mission phase, which was identified as being critical with respect to IMU gimbal lock limitations, was during the LM lunar landing. A hard over LM descent engine gimbal failure in the yaw direction would require positive pilot action almost immediately to avoid gimbal lock. From a vertical hover orientation, a random tumble from the vertical has an 11% chance of passing within 10 degrees of the lock orientation if not stopped before 90 degrees. If the attitude information was lost by such a maneuver, the vehicle would be thrusting nearly horizontally and probably downward unless the engine was immediately shut down which, of course, would be the proper first action during this emergency.

### 3.1.3 Considerations of a Four Degree-of-Freedom IMU

Difficulties near gimbal lock can be avoided by the addition of a fourth gimbal to the IMU. This additional gimbal is called the redundant gimbal since it provides more degrees of freedom than theoretically necessary on a geometric basis. This redundant gimbal, if used, would have been mounted outside the normal outer gimbal. The order would have been inner, middle, outer, and redundant. See Figure 3-3. The most likely operation would use the inner three gimbals to drive the stabilizing gyro error signals to zero while the fourth gimbal would be driven so as to keep the middle gimbal near zero and away from the gimbal lock orientation. This could be done by generating a redundant gimbal rate command by expressions similar to

$$A_{(redundant)} = \frac{k \sin A_{(middle)}}{\cos A_{(outer)}}$$

so that a negative feedback occurs to drive the middle angle towards zero.[1] It should be possible to make the inner three gimbals have the same dynamic performance as the simpler three degree-of-freedom system. Any base motion coupling, though, avoided in the Apollo system as described earlier, would make redundant gimbal motions appear as disturbances on the middle or inner servos, thus requiring special attention in loop responses. The redundant gimbal must be accelerated if it is to do its job, even when the middle gimbal angle is near zero. In fact, a situation like gimbal lock occurs for outer angles near 90 degrees as can be seen in the above equation. Close to 90 degrees outer gimbal angle, the redundant gimbal must be driven at a very fast rate to hold the middle gimbal angle at zero. In practice, this offers no real difficulty as long as the vehicle body rates are within certain limits.

With the four degree-of-freedom gimbal system, there are no constraints on vehicle attitude although rate limits do still exist for certain paths of attitude motion.

---

[1] A Nonlocking Four-gimbal Method of Isolating a Platform From a Rotating Vehicle, by Richard C. Hutchinson, MIT/IL Report R-285, April 1961, Chapter 8.

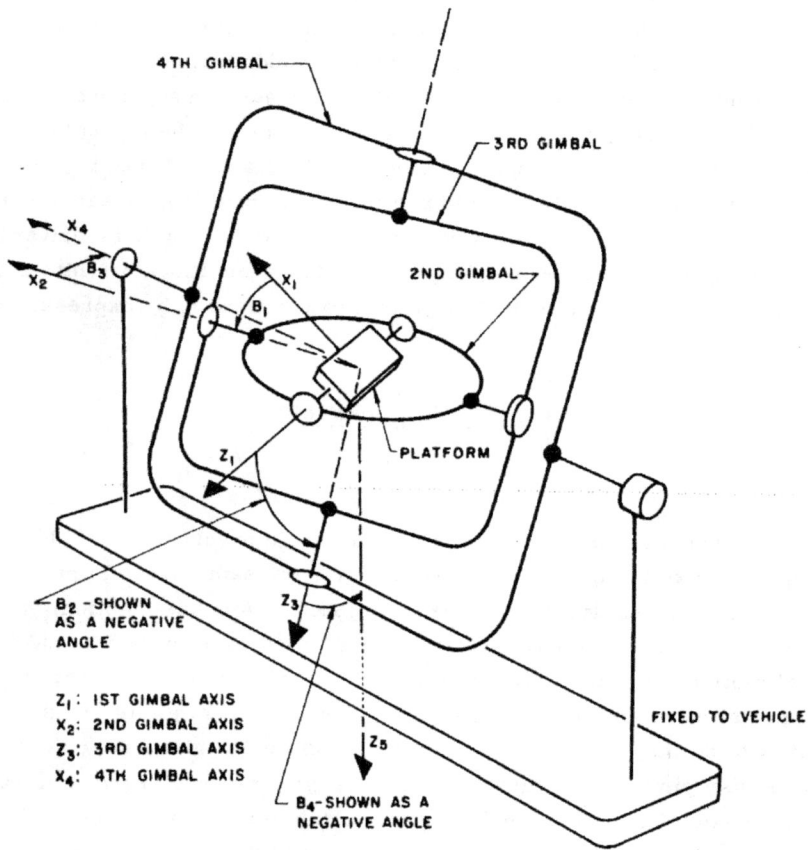

4TH GIMBAL

3RD GIMBAL

$X_4$

$B_3$

$X_2$

2ND GIMBAL

$X_1$

$B_1$

PLATFORM

$Z_1$

$B_2$ - SHOWN AS A NEGATIVE ANGLE

$Z_3$

$Z_1$: 1ST GIMBAL AXIS
$X_2$: 2ND GIMBAL AXIS
$Z_3$: 3RD GIMBAL AXIS
$X_4$: 4TH GIMBAL AXIS

FIXED TO VEHICLE

$Z_5$

$B_4$ - SHOWN AS A NEGATIVE ANGLE

Fig. 3-3   Four Degree-of-Freedom IMU

The disadvantages of using a four degree-of-freedom IMU were considered significant. To execute the fourth gimbal would have required an extra servo different from the other three servos in dynamic design. Extra data transducers, or the equivalent, on the middle and outer axes to determine the redundant gimbal commands would have also been required.

The expense involved in having to add the fourth gimbal depended upon a number of factors. In order to achieve the required guidance system performance demanded of the lunar mission, the desired IMU alignment accuracy stood out as a critical design factor. This meant gimbal data transducers having peak errors no worse than 20 seconds of arc would be required in order to use the optics system for star alignment references. The addition of a fourth gimbal to achieve the alignment would have compounded the problem in the IMU structure, in the computer interface, and in the flight computer program.

At the time, the 20 arc second accuracy data transducers determined the size of the Apollo IMU. Without these large units the IMU would have been several inches smaller in diameter. It was estimated that a fourth gimbal using these same proven transducers would have raised IMU structure weight by 15 pounds and increased the volume by 725 cubic inches.

The Polaris advanced guidance system in a four degree-of-freedom configuration was estimated to weigh only 2/3 the Apollo three degree-of-freedom unit and would have had equivalent performance except for the accuracy of the axis data transducers. Advances in small accurate data transducers were being studied with the goal of direct digital encoding of gimbal angles. This could have led to simplification of the CDU's, but the interface with the existing "eight ball" attitude display would have been much more difficult.

Additional complexity would have resulted in making full use of a four degree-of-freedom system. An additional CDU (3 pounds plus electronics for Block I hardware) would be required for flexible operation with the computer. The additional gimbal would have required a longer resolution chain to generate autopilot attitude error for pilot display. This chain was already difficult to achieve under the requirements imposed by the existing autopilot interface. Also, the computer would have had to assume the burden of providing the real time resolutions required to convert steering angles in accelerometer coordinates into proper CDU commands. Although this was done anyway in the more capable Block II flight computer, it was not realistic to do so in the Block I computer.

The addition of a fourth gimbal, other things being equal, would necessarily have increased the power drain on the main fuel cell batteries, besides having required an extra set of servo electronics. Heat transfer from the gyros inside to the heat sink on the housing would have been made much more difficult by the addition of the fourth gimbal. The redundant axis would have had to have slip rings to be useful and would have had to carry all signals on the outer axis rings plus those of the redundant axis instrumentation.

One problem considered was generating three degree-of-freedom attitude ball commands from a redundant four degree-of-freedom IMU. This could have been solved by the development of a compatible four degree-of-freedom ball, but the size penalty on the panel would have been prohibitive. The electrical resolution conversion from four degrees of freedom to the necessary three degrees becomes bound up in the equivalence of gimbal lock for which the redundant gimbal was to have prevented. Although the four degree-of-freedom IMU avoids gimbal lock and loss of attitude, the attitude information is difficult to display.

The advantages of the redundant gimbal seemed to have been outweighed by equipment simplicity, size advantages, and corresponding implied hardware reliability of the direct three degree-of-freedom unit. Since no operational difficulty of significance could be imagined with the three degree-of-freedom gimbal system, the decision to avoid the fourth gimbal was made.

In retrospect, some hesitation about the wisdom of the decision might be expressed. As stated earlier, the orbital operations during rendezvous with the "concentric flight plan" made the probability of desired orientations near gimbal lock much higher than the more efficient direct rendezvous originally envisioned where needed orbital plane changes were combined with the needed in-plane transfer orbit burns. There is no doubt that with present technology, a four degree-of-freedom all-attitude IMU could be produced with no more total weight, volume, power, or failure rate than the present three degree-of-freedom unit of 1964 vintage. But for that matter, the three degree-of-freedom unit would also correspondingly benefit from present technology.

### 3.1.4  Component Section

3.1.4.1  Gyroscopes.  The first step in selecting a gyro was to perform an error analysis of the expected trajectories to determine the criticality of performance. With the realignment capability and with position errors for landing of less than 10 nautical miles, a 10 meru[*] gyro appeared feasible. Several candidate gyros were considered. A modification of the type of gyro used for Polaris was chosen for the Apollo application. The production records were impressive and its performance was satisfactory. Moreover, the designers were familiar with the instrument. A decision was made to have separate NASA inertial component procurement and to have only one source for each component. The flight schedule was somewhat uncertain, however, and it was decided to allow as much time as possible for that contingency. The expected wheel hours in operation were to be about 2000 for early flight systems and about 1200 for later flights. This assumption has since been proven too low by a factor of about 50%.

Some design changes were made to the gyro which made it different from the Polaris instrument. The torque generator was changed from a two winding current product torque to a single winding current squared torque. This was done to provide the most efficient interface between the computer and the IMU, principally for alignment. The suspension frequency and signal generator frequency were changed from 4800 Hz and 800 Hz, respectively, to 3200 Hz. The change was dictated by the available computer timing pulses. 6400 Hz could have been selected but experience with higher frequencies was lacking.

To assure a quality gyro, acceptance specifications were tightened, and a reliability program was started which would provide reliability data and failure prediction methods applicable to production line gyros.

3.1.4.2 Accelerometers. The accelerometer selected was a modified Polaris 16 PIP. This selection was based primarily upon the simplicity of the PIP over the 16 PIGA (the other possible MIT/IL choice) and the inherent reliability believed possible. The performance requirements were not extensive, but were difficult for the PIPA to fulfill. Apollo asked for about 100 parts per million accuracy and an acceleration range of ± 16 g. The knowledge of torque generators and electronics being what it was dictated the choice of reducing the pendulosity from 1 gm cm to 1/4 gm cm, which had the obvious effect of magnifying bias instability. To achieve better scale factor stability, the tapered suspension was added to the design. A significant number of design changes were incorporated throughout the life of the program.

---

[*] 1 meru is a Milliearth Rate Unit equal to 0.015 degree per hour.

The initial conception for the accelerometer was that of a ternary or three state torqued accelerometer because the computer would not be able to handle the limit cycle operation of a binary ±3 Δ V pulse-input. It was further decided to have a three scale factor accelerometer. A nominal scale factor with a capability of about 16 g's was the primary mode. There was to be a low g capability where the Δ V scaling was to be decreased by a factor of four. This was primarily done to abet performance during the early phases of entry, particularly during the skip out portion. In the event of a steep entry provision, the accelerometer was subjected to the maximum acceleration capability by reverting it to a binary mode that increased the Δ V by a factor of approximately 2. These decisions, in view of the fact that many requirements were then unknown, represented the best engineering judgment available. Later, the accelerometer was changed to a binary torqued single scale factor device which permitted considerable system simplification.

3.1.4.3 Component Data Management. Both inertial components were to have common test equipment at the factory of the inertial component manufacturer, MIT and the Inertial Subsystem contractor. This communality of test equipment was in line with maintaining transfer of inertial performance from location to location. MIT thus maintained configuration control over both the PIP and IRIG test consoles. In the case of the gyro test console, the design work of the analog equipment was done by AC Electronics, and MIT designed the digital portion. MIT designed the final test consoles for the 16 PIP Mod D.

It was decided early in the program that MIT would maintain inertial performance parameters for all components. This early decision and its execution set up the now existing gyro and PIP performance parameter tabulation and distribution system. Initially, MIT set up the system and maintained records. Later, Dynamics Research was brought in to do the job which had grown considerably. The possibility now existed for comparing the Apollo gyro with the very similar Polaris gyro. Such comparisons were infrequent, however, and not of much value. Later, for other reasons, the entire job was transferred to AC Electronics. The data system improved with each transfer and, because of the renewed interest, is now a good up-to-date operating system.

3.1.5 IMU Design
The IMU stable member was made of beryllium to save weight and give better dimensional stability to the inertial component orientation. There were no gimbal stops on any axis, thus permitting any stable orientation. A decision

was made to provide a hermetically sealed IMU based on the requirement of operation in a vacuum. The enclosed air in the IMU was necessary to provide a medium for conducting heat from the stable member to the coolant.

Several components in the IMU such as torque motors, resolvers and slip rings needed further development. MIT developed and qualified two sources for each component; and AC Electronics, as the IMU manufacturer, made the independent supplier choice for production. The torque motor and slip ring developments were straightforward and proceeded smoothly. One slip ring vendor, however, was subsequently disqualified because of failure to deliver an adequate product.

3.1.5.1 IMU Angle Resolvers. The resolver development proceeded along the same general guidelines outlined above, but more elements, requiring more advanced state-of-the-art technology, needed developing. The resolver development was intimately tied to the method of angle interface between the IMU, optics, and computer. The method of angle interfacing with the CDU is described in Sec. 4.5. The system required 1 and 16 speed resolvers for the IMU, and size 8 and 11 one speed resolvers for the CDU. The required precision was about 20 sec for the 16 speed resolver. Development started on these configurations with two vendors. The competition engendered between the two vendors was very beneficial to the program, both from a technical development and a cost standpoint. There were at the start of the program several groups at MIT involved in resolver development, but later were centralized into one group for greater efficiency and competency. Later developments dictated the need for more resolvers for the sextant, scanning telescope and CDU. This development followed the pattern set initially.

3.1.5.2 Temperature Control. There was a serious attempt from the beginning to provide the temperature control system with all the reliability and flexibility of the original design requirements. Early attempts to use thermal heat of fusion salts were tried. This idea was sound but proved to be unworkable based upon mission time lines for IMU operation. The study was one of the first to definitize system operation on a lunar landing mission. The Block I configuration temperature control system provided a flexible, reliable system for manned flight. It was, in fact, over designed and, as such, was simplified considerably for Block II. Test and flight results to date have justified this simplification.

The thermal interface for the IMU and the G&N system in general was a difficult negotiation. It was decided early that the IMU should have integral cooling in the spacecraft. This decision was based on substantial technical data provided by MIT and proved to be a sound one inasmuch as no leak problems developed. Later, when integral cooling for all electronics packages in the G&N was to be actually implemented, it could not be negotiated. To provide integral coolant in the IMU, a roll bonding and passage inflation technique was decided upon. First attempts were promising, but the design was not ready for initial releases. A welded coolant passage was designed, but fabrication problems with the design lent impetus to the roll bonding technique which became incorporated at an early point in Block I production.

3.1.6 ISS System Design

Considerable effort was spent in making available as many backup modes as possible. Usage of segments of the system with other segments currently operating was a ground rule. The use of the IMU as an attitude reference without the computer was also incorporated.

An early attempt was made to put in manual $\Delta V$ steering by a visual monitoring of the Y and Z PIPA outputs. The astronaut would manually align the IMU with the X PIPA along the direction of thrust, then manually start and stop the engine, steer to maintain zero $\Delta V$ along Y and Z, and time the burn for the net $\Delta V$ gained. There were, however, operational problems with the design and associated with production. In the middle of Block I production, the above mentioned backup mode was dropped and the design was changed to reflect it. Other backup modes were retained but never utilized. All Block I flights were unmanned and no ability to utilize backup modes was available. The decision, at the advent of Block II, was to remove many of these modes of operation and concentrate on the primary mode of operation.

The test and checkout of the ISS were carefully controlled from a number of standpoints. Commonality of test equipment wherever possible was incorporated. Care was taken to have the system as close to its operating configuration as possible. It was not possible, however, to attain exactly that configuration. Later testing of the complete G&N system was to be in the spacecraft configuration, a decision which was difficult to implement in the beginning but was well worth the effort in the end.

In order to check for adequate mechanical and thermal design, electrical system functioning, and overall system performance, several IMUs were constructed, which included an engineering functional model and a production model.

Such multiple model construction permitted evaluation of various design phases to proceed simultaneously.

The performance requirements for the Inertial Subsystem and the G&N system were not clearly specified by NASA. The error analysis of the trajectories and early mission studies were conducted by MIT, and reasonable specifications were applied.

There was no target to hit as with an ICBM. For entry, the ships moved to pick up the spacecraft, or there was a little extra fuel to correct for steering errors resulting from inertial performance. This proved to be both a help and a hindrance. The hindrance, of course, was that MIT had to make decisions without all the program information available when necessary. It helped in that items could speedily be decided upon, and, of course, MIT gave more consideration to the detailed mission requirements. From an inertial performance standpoint, an analysis of the errors resulting from the IMU revealed that moderate performance could be required for manned missions. The most critical parameter was the gyro bias drift, and that was the result of the long time lapse between alignment and the thrust termination. Rather large errors in acceleration sensitive gyro drifts could easily be tolerated as well as moderately large PIPA scale factor errors. Moderate accelerometer bias errors could also be tolerated. A decision was made to hold a moderately tight inertial performance specification because of two factors: (1) the early flights were to be unmanned, and it was not known when nor what mission would be flown, and (2) the tighter performance was more conducive to higher reliability. Indeed, experience showed that on the unmanned flights, because of the variety of missions, a different error component was predominant for each unmanned mission. Many early decisions made during Block I production was possible based upon knowing the mission for which the system was scheduled. It was possible to avoid retrofit costs and speed up the deliverable hardware when the mission was known. For example, the backup $\Delta V$ lights were eliminated rather than retrofitted because of the unmanned first flight with the G&N. The requirement for maintaining the IMU at operating temperature all of the time was examined both early and late in the program. Because of the various operational and handling requirements, it was burdensome to keep the IMU always at operating temperature. There were two reasons for this requirement. The first was one of not knowing the performance effects of cooldown and warmup. Subsequent tests showed that the IMU could probably survive if the temperature did not go below 70°F. However, there were some detrimental effects from gas generation in PIPs as a result of lowered temperature, but these were corrected by subsequent design changes. Based on this reason alone, constant maintenance of IMU operating temperature seemed a good decision but no longer is considered so. The second reason for this requirement was the lower volume compensation limit of the PIP and IRIG bellows. Temperatures outside of the operating range tended to damage the component. This temperature limit for the PIP, the higher of the two, was about 65°F. A study of the environmental conditions showed that it was possible to go below this temperature. Since protection to stay above this temperature was more difficult to obtain than maintaining operating temperature, the latter method prevailed.

Selecting the basic timing signals to be available from the computer and the resultant frequencies of operation for the synchronized power supplies was an early fundamental decision; once made, changes could not be accepted without serious consequences. The decision was to have an 800 Hz gyro wheel; all else was to be a binary multiple of that. The suspension signals for the IRIG and PIP then would be either 3200 Hz or 6400 Hz as in the previous Air Force and Navy work. However, this decision in any future program should be carefully weighed prior to commitment.

The design and release to manufacture of the signal conditioners to be used for system checkout and telemetry considerably lagged the remainder of the G&N. This caused the late release of the harnesses and "work arounds" to compensate for lack of telemetry measurements during G&N factory checkout.

3.2    Block I ISS Development
3.2.1  General Chronology
During the Fall and Winter of 1961 - 1962, the ISS concepts described in Section 3.1.6 gradually evolved into design requirements. Preliminary design information from North American Aircraft Corporation (who was selected in November 1961 to design and build the Apollo Spacecraft) concerning interfaces and environment was made available. Material in this category was distributed in MIT/IL's in-house Apollo document, AGANI[*], during October, November, January, and February. At the same time, NASA's Langley Space Task Group was making available early reference trajectories for Saturn I flights SA-7 through SA-11. It is of interest to note that plans then called for MIT/IL's System No. 5 to fly as a passenger on Mission SA-9 in 1964. System No. 6, also in 1964, was to be flown on Mission SA-10 as a 7-day manned orbital flight.

---

[*]Apollo Guidance and Navigation Information.

Once the basic accelerometer and gyro designs were chosen and the three-gimbal IMU configuration decided upon, the design effort in specialized areas could proceed. Although much of the basic inertial subsystem was derived from the Polaris Guidance System design, the Apollo requirements differed from the very beginning because of the interface with the astronauts. Therefore, much conceptual work had to be done during 1962 to define the system modes and the display and control means available to the astronauts.

In the Block I system, the electromechanical CDUs were a basic element in displaying IMU gimbal angles to the crew and in commanding gimbal angles in a coarse align mode. Design work on the CDUs and the related mode switching resulted in initial definition in November 1962. At that time, the three ISS CDUs were still interchangeable with the two optics CDUs. Later on, the two designs became different because of special optics requirements and because the inflight repair policy was deleted.

The early ISS development models, in accordance with the development plan, included a Thermal Model, a Mechanical Integrity Model, a Vibration Model, and a Subsystem Prototype, ISS #4. Around this framework of developmental models, the design and fabrication effort moved forward during 1962. This section includes some of the highlights of the Block I design phase. An overall pattern of progress is visible from the general concept to the specific design.

In October 1961, the overall system was identified as having an inertial measuring unit (IMU), a sextant (SXT), an Apollo guidance computer (AGC), a power and servo assembly (PSA), a sun finder assembly (SFA), final approach equipment (FAE), an angular momentum wheel (AMW), and communications (CMM). The SFA, FAE, AMW, and CMM did not survive the initial system concept. In January of 1962, an initial procedure was identified for alignment of the IMU during prelaunch, inflight, and lunar countdown phases (at this early stage, the Apollo program had not yet settled on the lunar orbit rendezvous mission mode). The five Block I ISS modes of coarse align, fine align, CDU zero, attitude control and entry were well-defined in mid-1962, and in November of that year a first proposal for a switch configuration was published. Further consideration of switch design and moding considerations led to a final mode-switching circuit design in November 1963. The Block I ISS block diagram evolved gradually from an initial version in December of 1962 to a well-defined concept by May 1963. A master layout wiring diagram for the IMU was initiated in July 1962 and was the basis for IMU harness design, slip-ring wire assignment, and IMU connector selection.

The analytical work involving the relationship between inertial component performance and position and velocity dispersions could not get underway until NASA mission and trajectory designs were made firm. As early as July 1961, a preliminary estimate suggested that the then existing inertial state of the art would probably meet general Apollo mission requirements. Unfortunately, the initial pressure and ground rule did not permit careful consideration of technological gains in inertial components which were then available and could have been incorporated.

In November 1961, preliminary gyro performance specifications had been agreed upon. Actual error studies began early in 1962. The entry maneuver was studied first because its general nature was predefined and it would be the most critical for Apollo inertial subsystem performance. Results for this mission phase were published in June 1962 followed by a study of lunar orbit injection performance brought out in July. By early 1963, the Apollo mission definition was in a state which permitted analysis of other mission phases, such as translunar injection, lunar landing, and lunar take-off. These results were made available in March. At the same time, the errors involved in the process of fine aligning the IMU in space with the optics had been scrutinized. An error budget applicable to this process and including both IMU and optical errors was published in February 1963.

Consideration of the attitude reference system between the Guidance and Navigation system and the spacecraft began early in the program. Only in Block II did NASA call for direct control of reaction control system jets and the service module propulsion system by the G&N system, which then became the primary Navigation, Guidance and Control System or PNGCS. Actually, early in Block I there were some abortive studies of direct links between the G&N system and the spacecraft attitude and propulsion systems, but finally the Block I attitude error signals went only to the Command Module Stabilization and Control System (SCS).

By August 1962, this attitude error interface was defined as using the G&N carrier frequency of 800 Hz. The basic configuration of the resolver chain, which generated steering error signals from the IMU and CDU resolvers, was defined by September of that year. A preliminary design of the complete attitude error network was available in March 1963.

A specification of the IRIG performance requirements was identified as early as November 1961. The adoption of the Polaris 25 IRIG design for Apollo enabled specification work to begin on the pulse-torquing requirements which were crucial in the mission because of the multiple in-flight alignments. Specifications were firmed up for the pulse-torquing electronics by May of 1962. In March 1963, pulse-torquing tests of the Apollo IRIG indicated that there would be a problem of a bias shift depending on the sign (positive or negative) of the last pulse command. A requirement was generated for a computer pulse sequence that would always end with two positive pulses for bias reset, even for a negative command.

Turning now to a consideration of the IMU, the transition from the Polaris design to Apollo requirements proceeded quite rapidly. An inventory of required gimbal mounted electronics was published in December 1961 followed by an IMU wiring diagram in July 1962 and detailed specification of IMU/GME connectors in October of that year. Estimates of gimbal friction torque in January 1962 and of IMU gimbal moments of inertia in August helped with the design of the stabilization amplifier circuits. Bode plots of the stabilization loop dynamics were published in preliminary form in September 1962 and in final form, July 1963.

During the same period, negotiations were underway with vendors for the design, development and procurement of torque motors, single and multi-speed resolvers, and sliprings for the Apollo IMU. The procurement effort involved agreement on implementation of NASA's stringent quality assurance requirements for manned spaceflight hardware. In this connection, concern arose about the slip rings to be used in the IMU. These critical elements were extremely compact and yet had to carry 45 to 50 conductors in a noise-free, low-resistance manner. One vendor had a product submitted for Apollo consideration which had a history on other programs of unidentified contaminant growth. After review of the situation, this particular vendor's component was disqualified for Apollo use.

In December 1962, work on the temperature control of the Apollo IMU began with a general examination of the duty cycle and heat balance. The following April, specifications based on the IRIG and PIP requirements were made available to the temperature control designers. Electronic design of the redundant termperature control system then proceeded. There was a one-wire mechanization diagram available in October 1962, followed by the release of individual circuit design proposals and specifications the following January and February.

Packaging considerations for the ISS electronics, as contained in the PSA, had to follow the availability of overall circuit concepts. First estimates in this area were made available in March of 1962. In November of that year, a preliminary PSA envelope size was published along with a breakdown of this volume into eight "blocks". Concepts changed as more of the interface with the North

American Command Module was identified.  By April 1963, the Block I PSA had assumed its final form factor of 10 trays, which were removable in space for in-flight repair.

3.2.2  ISS Development Problems

Details about the early subsystem integration and test operations involving the Block I Inertial Subsystem are included in this section.  The material presented is indicative of the kind of problems that emerged during developmental testing. Although these specific problems were unique to the Apollo Block I ISS, their general nature provides worthwhile guidelines to other developmental programs.

A breadboard version of the GSE (Ground Support Equipment) was partially built by AC Electronics (then known as AC Spark Plug) in Milwaukee and given final form at the MIT Instrumentation Laboratory.  Test cables were constructed and a prototype computer simulator for providing clock pulse generation and output pulse shaping was fabricated by the Raytheon Company.

System testing of inertial subsystems was conducted to evaluate the fundamental soundness of the system design, evaluate preliminary test methods, train engineers, uncover problem areas, and evaluate design changes.  The following problems were encountered and corrective programs initiated:

(1)  Torque pulses in one Pulsed Integrating Pendulum Accelerometer (PIPA) loop were observed to couple into other PIPA loops.  This problem was substantially reduced by shielding the PIP torquer lines within the IMU between interaxis slip rings and by changing certain slip ring assignments.

(2)  The PIPA loop moding was found to be bistable, oscillating between its design value of 3:3 and a value of 4:4.  An increase in PIPA damping coefficient from 80,000 to 120,000 dyne-cm/rad/sec was initiated to insure solid 3:3 moding.

(3)  The 3200 Hz pulse width modulated temperature control system, carrying substantial power, was observed to generate 6400 Hz spikes which could be observed throughout the system.

(4)  The alignment and calibration of the gyros required modification of the original tests so that large error signals from the two gyros not under test would not couple through a common IRIG preamp output to bias the values measured on the particular gyro being tested.  Appropriate hardware changes were requested for

the production Apollo Ground Support Equipment (GSE) to provide gyro caging capability to maintain low gyro error signals on those instruments not under test.

(5) Attention was given to noise levels on the torque motor voltages. With the long ISS test cable runs, the levels were typically 10 to 12 Volts peak-to-peak. 300 k Hz oscillation on the IRIG error outputs was reduced to an acceptable low level by floating the individual shields along the interconnect cables to reduce existing cable line capacitance.

(6) Gimbal frequency response plots displayed secondary resonances by the middle and outer gimbals at 40 and 35 Hz which were not predicted by the linear transfer function for IRIG error response to a test input.

This transfer function had a first order lead term in the numerator at 161 rad/sec and a second order denominator term at 278 rad/sec with a damping ratio of 0.8. Although it is not now possible to determine the exact test conditions, these data do show an effect which, in Block II, was established as an OA coupling phenomenon.

(7) A synchronizing transient was found when going from the coarse align to fine align mode which is in excess of the expected transient resulting from coarse align loop standoff and IRIG float stops. This transient caused up to $10^\circ$ of gimbal angle offset and was attributed to a capacitor charge at the torque drive amplifier input of the servo amplifier. This capacitor charged up in the coarse align mode from the demodulated IRIG error signal and discharged upon entering the fine align mode. To prevent this large coarse-to-fine align synchronization transient from taking place, back-to-back diodes were used that dissipated the residual capacitor charge at the time of transfer.

(8) It was found that the middle and outer servo loops could be forced into a 6 Hz oscillation, depending upon the magnitude and phase of the IRIG error prevailing at synchronization. This problem was solved by the separation of power and signal grounds within the servo amplifier modules.

(9) In early closed loop PIP performance checks, it was discovered that the breadboard three axis loop, which was packaged in a single drawer, produced dc loop crosscoupling, which was almost entirely eliminated after "prototype" packaging was incorporated.

(10) The suspension power supply required a higher output voltage which was provided by increasing the output transformer turns ratio. This kept the suspension loop in the linear operating range after allowing for additional cable and slip ring resistance in the forward and feedback lines.

(11) Angular Differentiating Accelerometer (ADA)--The ADA is a damped torsional mass that senses inertial rotation under its own dynamics. The device has a low pendulosity, preferably zero. The problem cited arose during the acceleration phase of the IMU qualification program. The IMU was on a centrifuge mounted rigidly to the arm. During centrifuge testing, the gimbals oscillated with a frequency equal to the rotational frequency of the centrifuge. It became apparent that the ADA mounted to the gimbal was nonrotating and as such was under the influence of a rotating acceleration. With its pendulosity, the response was as if it were in inertial rotation with attendant stable member misalignments. A quick check revealed that there were no rotating accelerations for any missions that, with the specified allowable ADA pendulosities, would give any problem.

(12) Testing done in backup mode   showed that the suspension frequency was not held closely enough to maintain instrument centering as shown by excessive instrument quadrature levels and PIP timing pulses not compatible with loop operation. This backup mode was dropped from the system about February 1964 before further circuit evaluation was continued.

(13) Tuning capacitors were required at the encoder outputs to make them less sensitive to cable impedance matching effects which resulted in double-pulsing of the encoder.

(14) A great reduction of 6400 pps noise was attained by putting 0.1 uf across the temperature loop proportional control transistor and had the effect of increasing the temperature current rise time. The improvement in noise at the PSA was by a factor of six with little improvement found within the IMU.

(15) A capacitor was required across the suspension supply line at the stable member to tune the 2V 3200 Hz suspension input and permitted the suspension line to be phase-adjusted to the computer generated interrogate pulse train timing. This change resulted in more optimum timing within the torque-to-balance loops.

---

*The backup mode for the ISS assumed that the Apollo Guidance Computer had failed and all power supplies were free-running.

(16) A modified stable member harness featured shortened and twisted emergency and control heater lines, paralleled slip ring assignments on the ducosyn excitation, a changed suspension feedback location to permit better balance of the individual component suspension voltages, and PIP and IRIG torquer lines as twisted pairs. After an evaluation of these changes, a large reduction of the 6400 pps noise within the IMU was made by additional shielding incorporated at the intergimbal axis assemblies.

(17) An evaluation was made of additional stable member mounted heaters in which individual inertial component temperature gradients were measured with respect to the stable member orientation. First runs were made at high and low power conditions using the normal average IRIG sensor controller without the stable member heaters. Then both average IRIG and average PIP temperature control was evaluated at both extreme power conditions using the stable member heaters. The test results showed that by using the additional stable member heaters, average deviations of instrument temperatures versus stable member orientations were reduced from $1.5^\circ$F to less than $0.5^\circ$F. This testing was mainly concerned with the scale factor sensitivity of the PIPA loop to temperature, showing a difference between earth component calibration and the units performance in a zero gravity environment.

(18) The PIPA torquing loop was found to display an effect that caused the appearance of a deadzone about null during the issuance of torquing commands. The effect resulted from the interrogator which was a three state logic device with a binary switch maintaining the torquing current in the direction consistent with the last directional interrogator switch command. The problem arose when a small cross coupling of the interrogator switch output with the torquing pulse line occurred. Thus the resultant torquer current, applied during an interval when the interrogator was commanding, differed slightly from the torquer current applied during an interval that the interrogator was not commanding, but the binary switch was maintaining the direction of the last interrogator command. This effect called AKO (Alternating Kickoff) presented considerable problems in using the instrument for detecting the gravity perpendicular plane required for PIPA bias and alignment measurements. Design changes were made to correct this problem.

(19) An evaluation was made using a two level PIPA loop turn-on involving an initial low torquing current turn-on before the normal loop turn-on (See Section 3.3.2). This change had the effect of centering the float before applying full torquer current thus reducing the normal closure current hysteresis effects in the torquer until float-to-case alignment or torquer pole alignment was obtained.

Further, this change introduced significant PIPA loop bias performance improvement and was incorporated into Block I design.

By the spring of 1964, NASA's Apollo Program plans had matured, practical results from the Mercury program were available, and the final requirements on the PGNCS were being released. It was during this period that NASA and MIT/IL initiated discussions about a Block II version of the Apollo ISS. The genesis and direction of the Block II effort are discussed in Section 3.3.

This section completes a brief chronological review of Block I Apollo ISS design and development milestones. In March 1964, nearly all ISS production drawings, procurement specifications and test procedures were released. Block I entered its production phase during 1964 as AC Electronics Corporation and the MIT/IL design team turned primarily to the Block II design and development effort. However, continued support of the Apollo industrial contractors, Sperry, and AC Electronics was provided in the process manufacture, assembly, final test, reliability test, and qualification test. Similarly, as the Block I systems moved into an installation phase at the North American plant at Downey, California, and a checkout, countdown, and launch phase at the Kennedy Space Center, MIT/IL design staff members assumed temporary or long-term responsibilities in support of systems in the field.

3.3    Block II ISS Development

As the Apollo development became more advanced, a number of factors made a block change of design desirable. From the beginning, a block change concept was visualized as being inevitable, since the Block I design was created in the absence of many necessary guidelines and specifications. In July 1962, NASA changed the lunar landing concept from the earth orbital rendezvous to the lunar orbital rendezvous technique. Accordingly, in November 1962, Grumman Aircraft Engineering Corporation was selected to build the lunar module (LM). The MIT/IL system would provide navigation, guidance, and control. Thus, the LM concept made an obvious block change point for NASA's Apollo Program.

In June 1964, MIT/IL was asked to proceed with a Block II design for the command module as well as the LM. For both vehicles, the system was given direct interfaces with the gimballed primary propulsion systems, as well as the reaction control jet clusters.

3.3.1 Changes from Block I to Block II

3.3.1.1 <u>IMU Design</u>  MIT proceeded to make suggestions as to possibilities for
incorporating into the Block II package certain significant changes.  NASA at the
same time was viewing what they saw as possibilities for changes.  NASA asked
that MIT build a smaller, lighter IMU.  After studying the possibilities, MIT
recommended that by keeping the same stable member, the IMU could be reduced
by about 1/3 in weight with a corresponding reduction in diameter of 14 inches
to 12 inches.  The resolver chain simplification that came with the CDU changes
permitted a reduced number of resolvers.  The combining of 1 and 16 speed
resolvers onto the same rim reduced by three the number of resolvers, leaving
only one resolution and three angle measuring units in the IMU.  The removal
of three torque motors and the ADA's (Angular Differentiating Accelerometers)
and ADA amplifiers made possible the shrinkage of the intergimbal assemblies to
reduce the overall IMU weight and size.  The temperature control system was
simplified thus reducing the amount of gimbal-mounted electronics.  The IRIG was
to have more compact prealignment hardware by incorporating the preamp design
into the end cap hardware.  The PIP suspension module was redesigned to have an
integral assembly with a connector which would allow easy assembly of the PIP
into the IMU.

3.3.1.2 <u>PEA/PTA Design</u>  The LM installation presented significant problems for
the accelerometer in view of the seventeen feet of cable between the IMU and PSA
or prepared PIPA electronics location.  What was desired was a location close to
the IMU and on a cold plate with better temperature control.  Discussions with
GAEC revealed that it was possible to put the PIPA electronics in an assembly in
the vicinity of the IMU and also have its cold plate in series following the IMU,
thus achieving a lower cold plate temperature and a lower temperature deviation of
the heat sink.  The possibility of doing the same thing in the command module, was
of course, also considered.  The electronic  CDU was approved for incorporation
leaving the old CDU cold plate available which was in series right after the IMU in
the coolant loop and as such provided a colder, better controlled heat sink.  Since the
Block II design was to be humidity-proof, the concept of a sealed assembly was
introduced necessitating several changes in philosophy with respect to the
accelerometers.  There would be no further requirement for module interchangeability.  Because the sealed electronics package precluded final adjustment of
accelerometer bias and scale factor at the ISS or G&N level, the computer
compensation range was increased.

3.3.1.3 <u>Block II Gyro.</u> The most sensitive parameter for the gyro was bias drift. Several changes, which were aimed towards reducing this term, were incorporated into the Block II gyro design. Axial suspension was added, and radial suspension stiffeners were tripled to reduce float displacement with respect to the case. A reset winding was added to the torque generator, which, when excited with alternating current, was effective in reducing the residual magnetism of pulse torquing to a very low level. This residual effect was a source of inconvenience in the Block I instrument, for it required two additional pulses, a plus and a minus that provided no net torque on the float, but always left the torque in a standard state with a known fixed component of bias. Finally, a bias compensation winding was added to neutralize the net total fixed bias and to compensate for reaction torque changes resulting from voltage fluctuation of the signal generator and magnetic suspension excitation.

Other changes included a more efficient signal generator, and a redesign of the gyro prealignment hardware. The gyro wheel package was not changed.

3.3.1.4 <u>PSA Design.</u> The Block II PSA packaging was changed drastically from the Block I. All CDU electronics was removed from the PSA and much of the PIPA electronics. Since the Block II PSA was not constrained by the requirement for an in-flight module replacement capability, it was designed as a hermetically sealed assembly. The LM PSA was similar to the command module Block II PSA, except that it was smaller since it did not contain optical subsystem electronics.

3.3.1.5 <u>Block II CDU.</u> In 1963, an effort to replace the CDU gear boxes with an all electronics CDU was started. This resolver reading system was breadboarded and demonstrated to NASA. A weight estimate of about 18 pounds was given for the configuration devised to duplicate the electromechanical CDUs. This estimate was regarded as a significant weight savings. The system looked promising and NASA approved the concept for incorporation. Subsequently, a NASA decision to have a digital autopilot in both the LM and the CM resulted in numerous changes to the CDU. This and other decisions increased the CDU weight to its present 37 lbs.

3.3.1.6 <u>Moding Capabilities.</u> The new CDU and Apollo Guidance Computer (AGC) made numerous moding changes necessary and desirable. IMU Cage was to be the only manual mode except for system turn-on and coarse alignment. Everything else was to be moded by the AGC.

3.3.2   Block II ISS Development Problems

The breadboard Ground Support Equipment used for early Block I testing was modified, and additional test equipment and interconnect cabling were assembled to meet the new requirement of Block II testing.   The second GSE utilized was with Class A Ground Support Equipment which was furnished by AC Electronics and officially modified for the Block II and LM systems.

A breadboard LM ISS was assembled and used to perform a series of investigatory engineering tests.

Thermal tests were conducted to investigate required thermal padding methods and to determine the location of the stable member thermostat   such that the individual component had minimum sensitivities to gimbal position.

Experimental data of accelerometer bias changes were taken after closing the loops both from the null position and after storage in the plus and minus one gravity position.   Similar data were taken by closing the loop with reduced torquing current for a one minute period and then returning to the normal torquer current using the same closed loop positions.   Results of this testing showed that the low current closed loop position had improved bias stability.   This change was incorporated into both Block I and Block II PIPA loops.   The reason for this improvement is believed to have resulted from acquiring torquer pole alignment during the period of operation using reduced torquing current thus producing lower hysteresis effects.

To save IMU power throughout the translunar phase of the mission, the LM system was mechanized such that the suspension was shut off until 90 seconds before PIPA loop closure.   The effect of this turn-on was studied and it was determined that the sensitivity of PIPA loop bias to this type of turn-on was not significant.

Block II and LM G&N testing began in October 1965 and continued until July 1966.   Some of the more significant results of that testing are listed below:

(1) Expected high and low +28 Volt dc bus   ISS variation in the  LM  was 24.5 V dc to 32.5 V dc   at the Grumman G&N interface.   This bus tolerance compares to a 25.8V dc  to 31.8V dc ISS bus   voltage variation at the  C/M interface. Because the LM  buss tolerances were changed during the ISS design phase, and the voltage available at a particular power supply was about one volt lower than was

available at the G & N interface, the PIPA loop pulsed torque power supply (PTPS) required redesign to maintain voltage regulation over the limits of the expected LM bus tolerance. Loss of voltage regulation on the PTPS would have resulted in significant PIPA loop bias and scale factor performance degradation.

(2) The PIPA loop quadrature levels could not be verified in the system from PIP station calibration. Further checks showed that the stable member wiring and slip ring reassignments between the suspension lines and the instrument error signals reduced the quadrature levels in the system.

(3) The reset coil of the PIP was removed to reduce the effects on the deviation of scale factor as a result of excitation suspension overvoltage.

(4) During this period, PIP and IRIG performance data were accumulated. Alignment transfer measurements from the test area into the system were also made of the PIPs.

(5) A servo loop oscillation problem was found during coarse-to-fine align moding on the middle gimbal. It was later determined that loop oscillation could be induced on all axes with gyro float angles of from 100 sec to about 600 sec and, further, was sensitive to gyro temperature which caused a $5\%/^\circ F$ change in the gain of the transfer function. After the problem area was identified, it could be induced on all Block II systems tested.

To provide stable operation for large input signals, nonlinear compensation was provided by parallel signal diodes in the forward path of the servo amplifiers. The breakdown voltage of these diodes was about 400 mV. This voltage permitted the servo to be driven well into saturation before diode breakdown reduced the amplifier gain resulting in a sustained gimbal oscillation.

Bringing the breakdown level of the diodes down to 250 mV resulted in good loop performance at synchronization, yet the breakdown point was high enough to insure an adequate linear region of operation when considering both possible demodulator imbalance and the variations of the diode breakdown point with temperature.

(6) A comparison of the ability to obtain servo loop frequency response results using the normal GSE recorder to that of a Boonshaft and Fuchs servo analyzer was made. The results indicated a favorable comparison.

(7) Tests were performed which determined that IMU gimbal rates up to 12.5 radians per second could be attained from an IMU cage command. The rate limitation of the servo loop resulted from the torque motor back EMF.

(8) It was discovered that both the middle gimbal and outer gimbal torque motor voltage noise levels and the associated loop frequency response varied as a function of the position of the inner gimbal. A study of the gyro orientation and the gyro output equations, which were combined in the gyro error resolver on the inner gimbal, revealed that the error voltage applied to the middle gimbal and outer gimbal torque motors was a function of twice the inner gimbal angle. The other error term always present in the gyro output equation was the OA coupling term which will always appear and could cause this type of trouble in any system if careful consideration isn't given to gyro output axis inertia. To eliminate the sensitivity of the gyro loop transfer function gain to inner gimbal angle, the Z gyro was physically rotated about its output axis by $180^{\circ}$. Mechanically, this change necessitated installing a new set of alignment pins, displaced $180^{\circ}$ from the old ones, at the stable member. To maintain loop phasing, the Z gyro suspension voltage input was reversed, and the torquer lines were reversed at the Z gyro harness connector of the stable member.

(9) Problems were experienced with the IMU mounting bolts as a result of binding of the mounting screw at the IMU mounting hole. Alignment bushings were added to effect accurate IMU-to-navigation base alignment, and to preclude binding, sufficient clearance between the mounting bolts and the bushings was maintained.

3.4  Reliability
3.4.1  Scope

MIT organized an effective Apollo Reliability program that utilized accepted practices and controls to ensure that the reliability inherent in the system design was of the highest attainable. Furthermore, every effort was taken to assure that the reliability initially achieved would not become degraded during development and production. The first steps taken to provide NASA with system reliability commensurate with mission success and crew safety requirements was the preparation of the Guidance and Navigation Reliability Program (R-349). This document defined the specific levels of effort, MIT responsibilities, and the role of the participating contractors in the Apollo reliability program. In short, R-349 established the program in which MIT was responsible for administering and controlling the design, development, production, shipment and field use of G&N systems. It was shown therein that the participating contractor's function was to provide MIT with reliability support from their facilities within the scope of their negotiated contracts with NASA.

The scope of the G&N reliability program was established by NASA in NCP 250-1 (Reliability Program Provisions for Space System Contractors) and was implemented by MIT, the G&N industrial contractor, AC Electronics Division of General Motors and directed subcontractors, Raytheon Company and Kollsman Instrument Corporation. All aspects of this program were applicable to all elements of Apollo airborne hardware and mission essential ground support equipment.

3.4.2  Design Review

To assure that the potential for reliability existed in the design, and to achieve early maturity in design, considerable emphasis was placed on the design review aspect of the program. The Design Review Board (DRB) consisted of the G&N Technical Director and his designated design group heads. Each design group head in turn called upon design engineers and contractor personnel in his area of responsibility to cover the design being reviewed. The membership of the board was of personnel capable of providing a total assessment of the design under consideration. An MIT reliability group representative and a systems group representative were always present at DRB meetings. In addition to the technical aspects of the design, other items considered were interface compatibility, logistics, suitability for fabrication, assembly test or use, and reliability.

All documents describing or applicable to guidance and navigation hardware including layouts, engineering drawings, specification control drawings, procurement specifications, process specifications, assembly and test procedures, circuit schematics, and wiring and logic diagrams were subjected to a design review prior to the time of release or change. Essential reliability factors and principles of design considered in the Design Reviews were as follows:

(1) Failure modes, probable cause and effect.

(2) Protection against human initiated failure or induced failure from other causes.

(3) Standardization of design and parts.

(4) The elimination of inherently unreliable components.

(5) Use of known materials and processes.

(6) Component part application and derating.

(7) Producibility

(8) Maintainability

(9) Simplicity

(10) Safety and human engineering.

(11) Conformity to environmental specifications.

The complete details and operating procedures of the MIT Design Review Board are contained in MIT Report R-496 (Design Review Procedures). Included therein are mechanical and electronic design review check list forms, electronic component stress analysis form, and DRB report form.

The contractors augmented the MIT effort described above by reviewing all designs for considerations of producibility, optimization, standardization and additional reliability of quality assurance and elements. They conducted design reviews to great depth on elements of the system for which they were delegated design responsibility. Recommendations stemming from contractors' reviews were processed by them through the MIT engineering groups and then through the MIT DRB.

All design releases, drawings, specifications, and documents associated with manufacture fabrication, and test of Apollo Guidance and Navigation equipment, and all changes that were made on these documents issued from the formal procedures of the Change Control Board. Configuration management, control and identification requirements were defined by NASA in document NPC 500-1. The procedures used by MIT to implement those requirements were defined in MIT report E-1087.

3.4.3 Reliability Testing

The design environment within which the Guidance and Navigation was required to operate was defined in Interface Control Documents (ICD) negotiated between MIT and the spacecraft contractors, North American Rockwell Corporation and Grumman Aircraft and Engineering Corporation. North American was responsible for the Command Module and Grumman Aircraft & Engineering for the Lunar Module. The design environments included such things as acceleration, vibration, shock, temperatures, humidity, pure oxygen environment, electrical input power, and pressures. Since these were negotiated early in the program when the anticipated environments were largely unknown, the ICD design limits were generally conservative.

Design evaluation testing was conducted early in the design phase on mockups, prototypes and first article development hardware to assure that the equipment as designed did indeed possess the integrity and have the capability to meet and exceed performance requirements, and to determine and define margins and limitations of the design in excess of requirements. The design of each element was rigorously examined with regard to thermal evaluation, mechanical integrity, marginal voltages, vacuum, functional and operating characteristics, stability, alignment, system integration and interface, and other peculiar characteristics or environments such as humidity, salt, contaminants, and electromagnetic interference to which a particular element is sensitive.

A formal Qualification Test Program was established to provide maximum assurance that the G&N equipment performed its required functions under the environmental conditions for the Apollo Mission. ND 1002037 (Apollo Airborne Guidance and Navigation Qualification Specification) identifies the elements of the G&N system and the block configuration to be qualified to each type of environmental stress level. In general, the total G&N system was qualified to nominal mission levels, and the subsystems and subassemblies were qualified to design levels with overstress in critical environments. Parts were qualified to a design level with emphasis on ability to determine part quality. The qualification criteria for parts were established by: (1) the expected maximum stress level anticipated in the worst case system application, (2) and adequate margin of safety, and (3) the degree to which a measure of quality in the manufacturing techniques was desired.

Separate reliability programs were conducted for the IRIG and PIP. They are discussed in the IRIG and PIP sections of this survey.

### 3.4.4 Reliability Analysis

During the conceptual phases of the MIT design effort, considerable reliability analyses were performed on the Apollo mission and various design approaches or system configurations to estimate mission success probabilities. Techniques of redundant computers, IMUs, and inflight maintenance were considered. As a result of these studies, it became readily apparent that mission reliability requirements for guidance could be attained without either major redundancy within the system or inflight maintenance if the reliability of two system elements could be significantly improved over current experience, namely, the computer micrologic and the gyros used in the IMU.

The micrologic problem was attacked by a searching test activity leading to an understanding of the various failure mechanisms prevalent in the devices. Correction methods and control of supplier processes incidental to the fabrication of these devices were instituted followed with special screening tests on a lot basis where dropouts were classified and special criteria imposed to allow judgment on the acceptability of each lot for Apollo flight hardware.

MIT's analysis of the Apollo mission and experience with gyros in ballistic missile guidance systems clearly indicated a positive technique of assuring that each flight gyro would have an MTBF inflight in excess of 100,000 hours. An exhaustive test program was conducted to study the effects of turn-on and turn-off versus continuous running on performance parameter degradation and, in addition, develop a technique of predicting impending failures. Using this test data as a basis for selection, the performance data on the total population of Apollo gyros were evaluated and criteria evolved for comparing the performance history of a specific instrument against the entire population thus allowing a highly confident assessment of a specific gyro's reliability over the next several hundred hours of operation. Such an assessment was made periodically on each gyro in service and shortly before committing any gyro to a flight mission.

### 3.4.5 Materials

During the hardware design phase, MIT established a "Material Review Board" very similar in nature to the Design Review Board previously mentioned.

This board, comprised of qualified engineers, chemists, metallurgists, and reliability experts, functioned to review all materials selected for use in the guidance system design. All material was considered for its potential effect upon the vehicle environment, i.e., outgassing, toxicity, flammability, combustibility, and interface compatibility with other materials. Areas of question as to suitability in the design were resolved by test and experiment. As a result, materials having undesirable characteristics were eliminated from the design, thus generating the highest confidence in acceptability of remaining materials.

3.4.6 Component Parts

During initial design stages of the guidance system, MIT employed a specialized team of reliability engineers to work directly with design engineers in the selection and application of parts. Parts were procured only from approved and qualified vendors. Throughout the Apollo program, all sources of supply for parts were strictly controlled by MIT. New parts as required were extensively evaluated and ultimately qualified for Apollo usage. This effort included the preparation and release of very strict specifications of each part for use in fabrication of flight systems.

In addition, MIT established and implemented, with the support of participating contractors, a system for controlling the processes and fabrication techniques employed by suppliers of all parts having a critical application within the Apollo guidance system. This effort, which was largely successful, has been responsible for assuring uniform quality levels of purchased piece parts and components procured from time to time throughout the duration of the Apollo program.

There were two basic sources of information to assist the designer in the selection of a purchased part. The first was the Qualification Status List, (QSL), ND 1002034 which provided part identification and status of qualification.

The second source of information was the Standard Parts Manual. This manual contained the procurement specification control drawings (SCD) on all parts listed in the Qualification Status List as well as other nonpreferred items that were being used on other projects. Before any nonpreferred part could be used, it had to be added to the project QSL.

In the event a designer could not find an appropriate part, desired additional interpretation on information found in the two documents mentioned above,

or wished to use a nonpreferred part, he had to seek assistance from the
Reliability Parts and Materials Group. This group consisted of parts specialists
whose job was to control the qualified suppliers list and be aware of the capability,
quality and availability of parts in their particular field. This awareness included
"state of the art" exotic devices as well as time-honored, standard military parts.
An individual of this group worked with the designer in the identification and
selection of the required parts and many times could offer assistance in vendor
contact and expedite sample procurement. This same individual, upon agreement
of the designers that the part was indeed suitable, would initiate the request for
documentation.

The program Quality Assurance requirements for suppliers of parts and
materials was delineated in Apollo G&N document, 1015404, which imposed rigid
configuration, process and Quality Control requirements. These requirements
were divided into three categories: class 1 (critical), class 2 (semi-critical), and
class 3 (noncritical). Class 1 was invoked for parts that contributed greatly to
mission reliability or had widespread usage. Class 2 was specified for parts that
were potential contributors to mission failure but were not used extensively. Class
3 was usually reserved for noncritical mechanical hardware items. Documented
and enforced supplier control was required for parts in class 1 and 2, which had to
be negotiated. The contractors implemented this aspect of the program and assumed
the responsibility for monitoring supplier quality by means of surveys, inspection
data and documentation control. Included in documentation control was traceability,
material lot and/or age control, and process controls. The category of each purchased item was identified on its specification control drawing.

3.4.7 Failure Reporting and Corrective Action
The Apollo Guidance and Navigation Failure Reporting and Corrective
Action System is described in great detail, complete with forms and implementation
procedures in MIT report E-1322. For this reason, just a short summarization
with an accompanying flow chart is presented in this document. The requirements
of E-1322 were effective for MIT and each of the contractors beginning with
acceptance tests at the assembly level and continuing throughout equipment life.
Regardless of location, the current failure reports were given appropriate
dissemination to MIT and the contractor whose hardware may be involved. The
original copies of the failure reports were collected and correlated at the Guidance
and Navigation Data Center. Failure analysis and corrective action responsibilities
were assigned by the respective contractors and effectiveness monitored by cross
correlation to individual failure reports. Any item that had not received appropriate

analysis or corrective action remained a problem area and was periodically reviewed. E-1322 also contained information relative to how the accumulated data were handled and analyzed in order to present meaningful failure information and trend indications.

As the flow chart in Fig. 3-4 also indicates, failures or discrepancies occurring during in-process fabrication, assembly, or test within the contractors' organizations were handled with the individual contractor's internal reliability procedures unless they were of sufficient import to warrant consideration under the plan described in this document. The detailed implementation procedures and a definition of this activity were contained in the contractor's quality plans.

Failures occurring during the design and development phase at MIT were reported to the MIT reliability group on the short form (see Fig. 3-5) for cataloging and following proper corrective action. Information pertaining to significant development failures was transposed by the Reliability Group to the G&N failure form as required by the G&N failure reporting and corrective action system (E-1322) and transmitted to the G&N data center for processing as part of that system. As a result, all significant failures occurring during design verification tests, qualification tests, and tests on deliverable end items or spares were included in this system. Not necessarily included, unless of special importance, were data from breadboard evaluation, and part and material design evaluation.

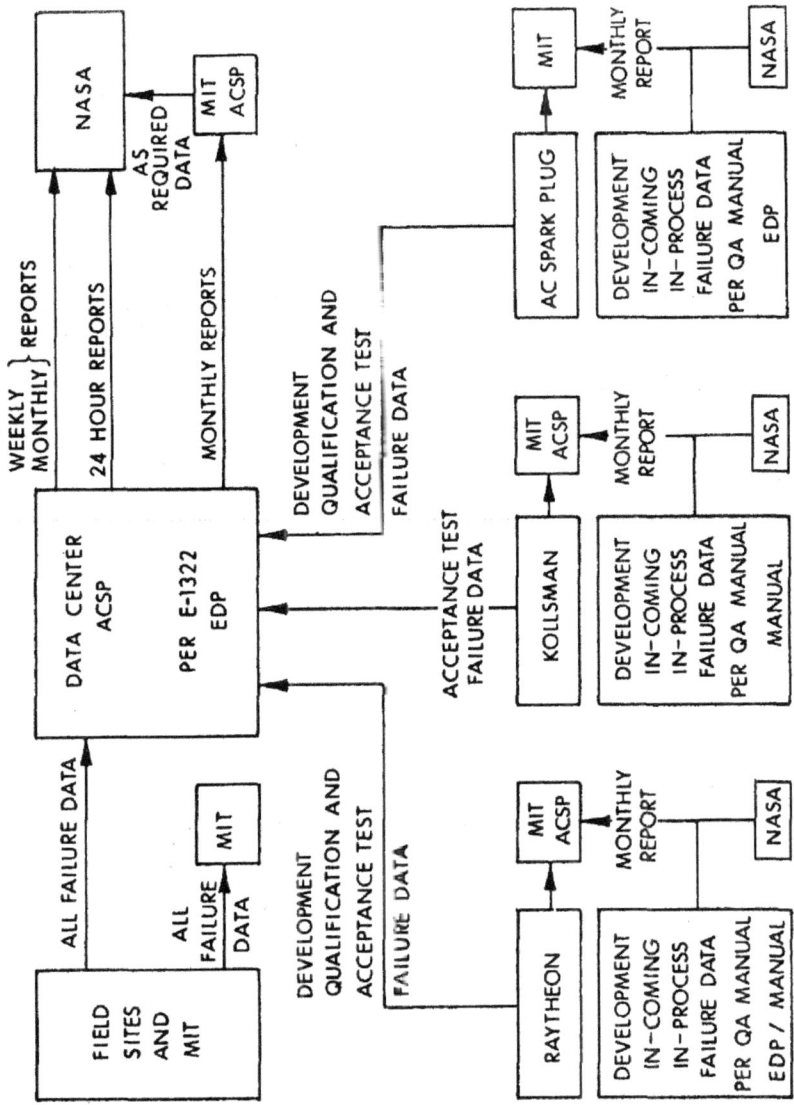

Fig. 3-4 G&N Failure Data System

MASSACHUSETTS INSTITUTE OF TECHNOLOGY
APOLLO GUIDANCE AND NAVIGATION
MALFUNCTION REPORT FORM

| Sub-assembly Name | Serial No. | Location | Date |
|---|---|---|---|
| Failed Part | Value | Manufacturer | Mfg Part No. |
| Part Operate Hours | Description of Oper. Conditions | | |

Description of Trouble:




Cause:


Action Taken:

Submitted by _____

Group _____


Figure 3-5

Section IV

INERTIAL SUBSYSTEM (ISS) COMPONENTS

## 4.1    Pulsed Integrating Pendulous Accelerometer (PIPA)

4.1.1   General Description.

The PIPA (Pulsed Integrating Pendulous Accelerometer) used in the Apollo
Inertial Measurement Unit was a single degree-of-freedom specific force integrating
receiver, which operated closed loop as a torque-restrained pendulum.  The instrument consisted of a pendulous float magnetically suspended in a viscous fluid, signal generator and torque generator microsyns, and associated electronics and calibration modules.  Figure 4-1 is a schematic diagram of the Pulsed Integrating
Pendulum (PIP).

The pendulous float was a hollow beryllium cylinder with built-in mass imbalance.  Ferrite rotors were mounted at each end of the float and supported the magnetic suspension and microsyn functions.  The float was surrounded by heavy fluid
that provided both a buoyant support and viscous damping.

An applied specific force along the input axis of the pendulum caused a rotation about the output axis.  This rotation was sensed by the signal generator and
used as an error signal to initiate a response from the torque generator.

A cutaway view of the complete pendulum is shown in Figure 4-2.  The pendulous mass was mounted on the periphery of the float and extended into a small
groove in the damping block.  The width of the groove allowed a maximum float
rotation of $\pm 1^{\circ}$ of arc.  The magnetic suspension stators were positioned opposite
the inside portions of the rotors and formed part of a coaxial structure with the
torque generator on one end of the instrument and the signal generator on the opposite end.  Four bellows provided for volumetric compensation of the damping fluid
between $10^{\circ}C$ - $71^{\circ}C$.  Two alignment rings on the end of the instrument permitted
accurate positioning of the input axis of the PIPA relative to the mounting surface.
The instrument was aligned to the mounting ring before incorporation in the IMU.

The overall block diagram showing the pendulum within its control loop is
presented in Figure 4-3.  The signal from the signal generator was amplified and
used as an input to a sampler (called the interrogator) which sampled the signal at
discrete times determined by the guidance computer.  If the signal exceeded either

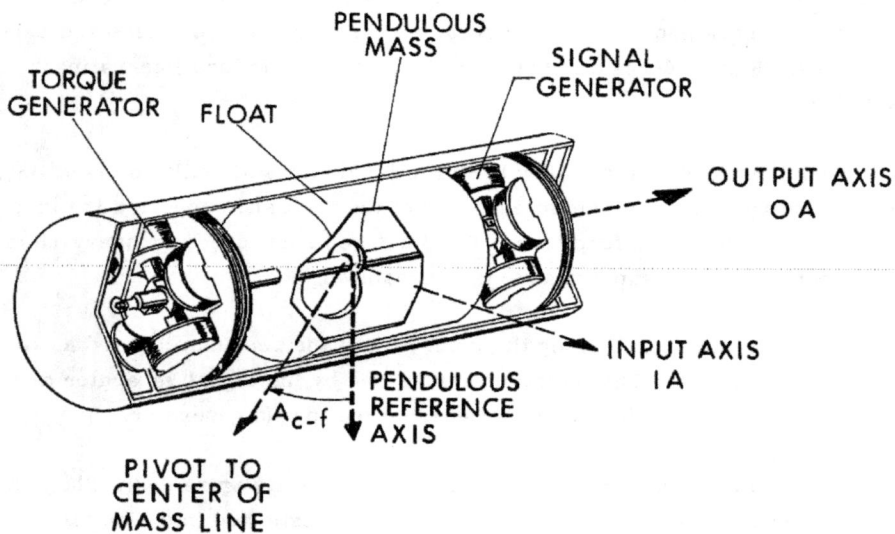

Fig. 4-1   Pulsed Integrating Pendulum Schematic

Fig. 4-2  16 Pulsed Integrating Pendulum, Mod D

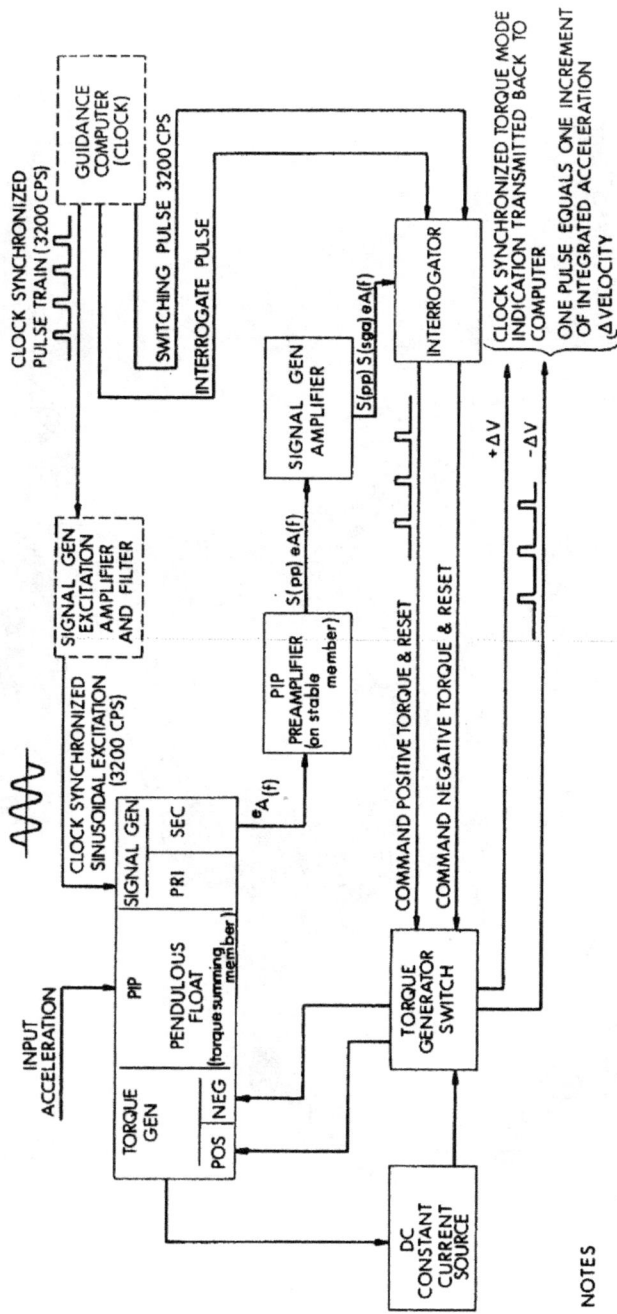

Fig. 4-3  PIP Accelerometer Block Diagram

the positive or negative threshold of the interrogator, a command pulse was sent to the torque generator switch. The switch, in turn, commanded the proper polarity current to the torque generator of the instrument and nulled out the error signal from the signal generator. The switching times were precisely controlled by the switching pulse train. The torquer current magnitude was controlled by an ultra-stable dc current loop. Torque was applied to the float in discrete increments, and for each increment, a pulse was generated and sent to the computer to provide a continuous monitoring of the total torque applied to the instruments.

### 4.1.2 Operation.

4.1.2.1 Magnetic Suspension. The FIPA made use of two quasi-elastic magnetic supports to properly position the neutrally buoyant float within the case. A cross section of the magnetic structure used is shown in Figure 4-4. This figure depicts the unique feature of separated suspension and transducer functions using a single coaxial structure. Radial and axial centering was produced by action of the suspension magnetic field on the inner surface of the tapered rotor.

4.1.2.2 Signal Generator. Float rotations were sensed by an angle-to-voltage transducer called a signal generator. A cross sectional view of the signal generator stator and rotor is shown in Figure 4-5. The primary coil of the signal generator (SG) was wound in series around each of the eight stator poles. The flux paths set up by these poles were typified by the flux paths shown in the figure. The SG secondary coils were wound such that the even-numbered poles were 180° out of phase. If the contoured outside diameter of the rotor was perfectly centered and at zero rotation relative to the stator (as shown in Figure 4-5), the total signal induced in the secondary of the SG would be zero. If a rotor (float) rotation occurred in either the positive or negative direction, a corresponding in-phase or out-of-phase output was obtained across the SG secondary, since the reluctance seen by the positive secondary poles was no longer equal to that seen by the negative poles.

4.1.2.3 Torque Generator. Except for the way the coils were wound, the torque generator (TG) cross section was identical to that of the SG. The position of the TG coils and resultant flux paths for positive torque are shown in Figure 4-6. Since only one torquer was on at any time, there were two minimum-energy rotor positions: one for positive torque, and one for negative torque. The rotor always tended to line up at the minimum-energy position opposite the working poles. However, the frequency of switching allowed only a minute oscillation about a center point midway between the minimum energy points. The torque produced was proportional to the square of the flux linking the rotor poles and thus was a function of the square of the torquer current.

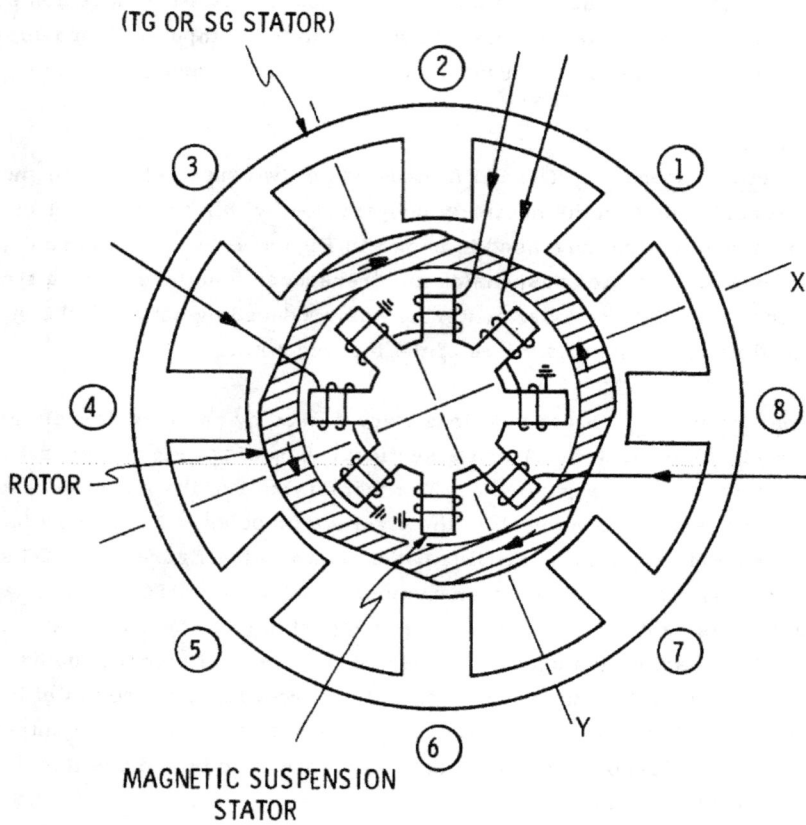

Fig. 4-4  Apollo PIP Magnetic Suspension Cross Section

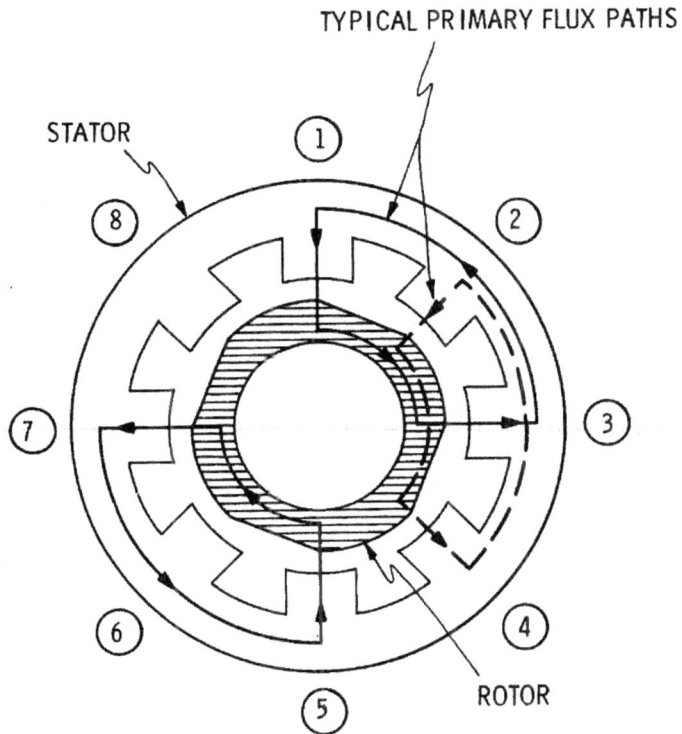

Fig. 4-5  Apollo PIP Signal Generator Cross Section

TORQUE CURRENT
RETURN

TORQUE CURRENT

TORQUE CURRENT

STATOR

ROTOR

FLUX PATHS FOR POSITIVE
TORQUE

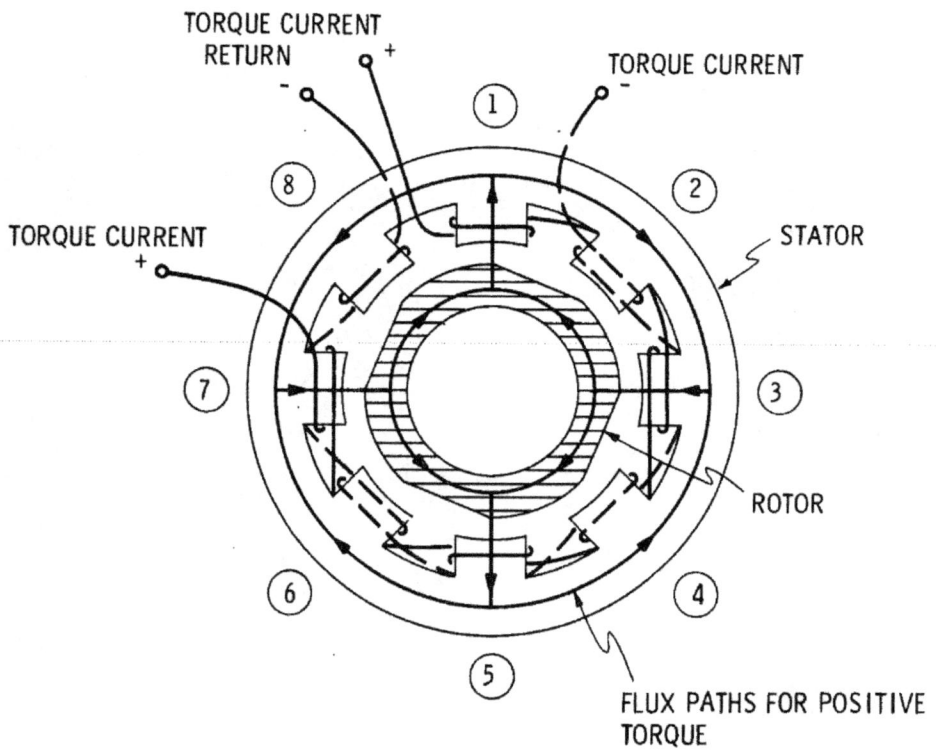

Fig. 4-6 Apollo PIP Torque Generator

The resulting torquing loop was a two-state or binary loop with torque applied continuously in either positive or negative directions. The mode of oscillation of the float was determined by the design parameters of the instrument. If m pulses occurred alternately in each direction, an m:m mode was declared to exist.

The velocity equivalent of one torque pulse was defined as the scale factor of the instrument having the units of cm/sec/pulse. Inequalities in the positive and negative torquing circuits caused an imbalance in the resulting output pulse train (with no input). The measure of this imbalance was defined as the bias. Bias was usually defined in terms of an equivalent acceleration applied to the float; e.g., a bias of 1 dyne cm is equivalent to an acceleration of $4$ cm/$\sec^2$ applied to a float having a pendulosity of 0.25 gm cm.

4.1.2.4 Ducosyn Flux Patterns. Figure 4-7 shows the flux pattern present in the ducosyn. The heavy lines represent the flux paths for the odd poles (+ torque) and the dashed lines represent the flux paths for the even poles (- torque). Also shown are the flux paths for the magnetic suspension. Since the rotor was a single ferrite piece, these fluxes were all superimposed in the rotor. The single piece ferrite rotor used in the Apollo PIP allowed interaction of the suspension and torquer fluxes. Assymmetries in the superposition of these fluxes in the rotor could have caused variations in PIPA bias which were dependent upon the phase relationship of the fluxes. Operation in the reverse torquer configuration (reversal of current flow in one torque winding) brought about a symmetric superposition of fluxes which would be almost independent of phase and would have reduced any bias changes to second order.

Three different conditions of flux were present in the stator. The flux existing in the poles was the on-off type, as shown. The flux in the back iron section, b, d, f and h, reached a dc level once the unit was turned on. (This condition remained constant except for a slight transient that occurred at switching intervals.) The flux in sections a, c, e and g was an alternating one. The stator flux magnitude should not have changed regardless of the torquer winding that was being energized; only the duty cycle of the positive and negative pulses should have changed. Changes in the permeability of the ducosyn materials would cause variations in the flux magnitude resulting in changes to the scale factor and bias. Permeability changes could have occurred as a result of transient overshoots in the torquer or suspension currents at system turn-on and by stresses incurred as a result of slow curing of microsyn encapsulating materials.

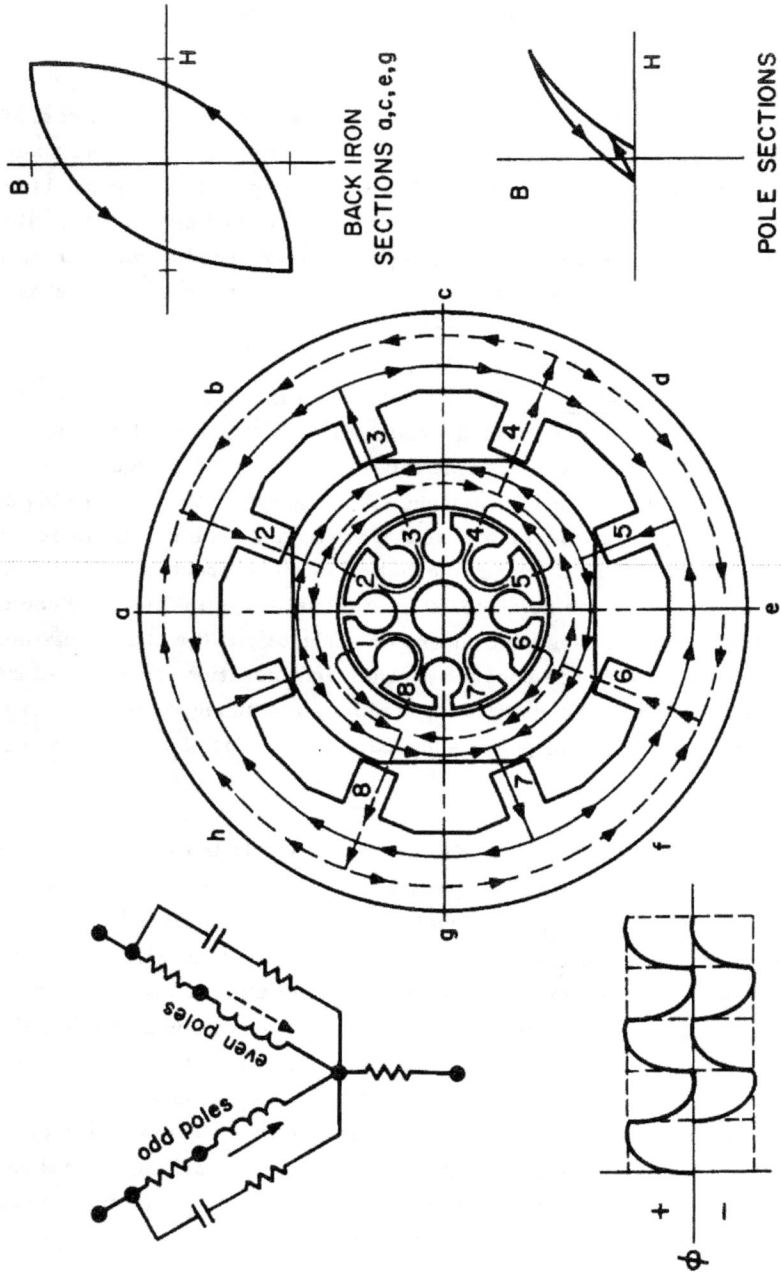

Fig. 4-7 PIP Ducosyn Torquer Magnetic Paths

BACK IRON
SECTIONS a, c, e, g

POLE SECTIONS

even poles

odd poles

4.1.2.5 Electronics. A functional description of the electronics involved with PIPA operation is depicted in the control-loop block diagram, Figure 4-3. The filtered SG signal was amplified by the preamplifier and ac differential amplifier and sent to the interrogator. The interrogator utilized a dual channel peak detector that was synchronized by the interrogator pulse train from the computer or system clock to determine the polarity of the signal from the ac amplifier. The outputs from the peak detector controlled the state of two multivibrators that commanded either positive or negative torque from the switch.

The binary current switch controlled the direction of the current flow from the current source in accordance with the commands received from the interrogator. A switch pulse-train synchronized the switching of the second multivibrator stage in the interrogator such that the pulses of current sent through the torque windings were of equal width.

The current loop is expanded in Figure 4-8. The output of the switch was sent to the calibration module. This module provided the necessary passive components to make the torquer coils of the PIP appear purely resistive thus controlling switching transients and optimizing current stability. Bias adjustment was obtained by use of the resistive shunting circuit. Fine padding of the resistive loads, $R_1$, in series with the torque windings, allowed equalization of the time constants, $T_1$ and $T_2$.

The voltage drop across the scale factor (SF) resistor was compared with a precision voltage reference (PVR). The error voltage was amplified by a high-gain dc differential amplifier and used to control the current magnitude in the loop.

4.1.2.6 Mode of Operation. Initially, the Block I Apollo torquing loop was designed as a three-state or ternary loop with no torquing pulses applied to the float if the output of the ac amplifier was less than the interrogator threshold. This approach was taken to conserve spacecraft power. Testing of the early PIPAs revealed that, as a result of the magnetic hysteresis of the torquer stator and rotor material, a significant residual flux remained in the core when the torquing current was removed and caused the instrument to show large bias variations that were dependent on the polarity of the last applied torquing pulse. It was further determined that the residual effects could be largely eliminated if alternate torquing were imposed continuously on the torquers thus creating a nearly constant average magnetic state in the materials. Consequently, a decision was made to establish a binary loop to neutralize the effects of the hysteresis.

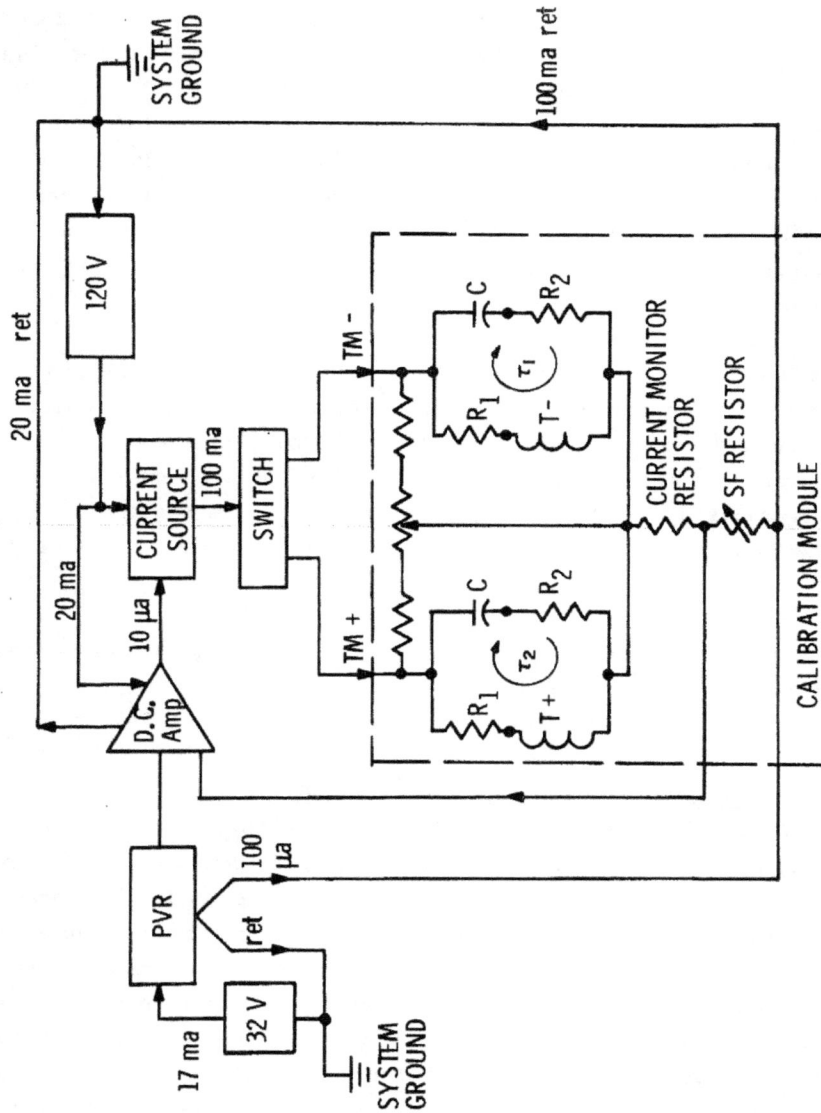

Fig. 4-8 Apollo PIPA dc Current Loop

4.1.3  Problems and Solutions.

4.1.3.1 Scale Factor Instabilities. (1) Torque Current Overshoot -- Torque current transients in excess of normal operating current levels will change the permeability of ducosyn magnetic materials by raising the operating level on the B-H curve. This phenomenon was identified early in the program. It resulted from the finite time required for the torquing electronics to stabilize during turn-on and also because of the decay characteristics of the electronics during turn-off. Design changes were made in the electronics which limited these transients to acceptable levels.

(2)  Excitation Overshoot -- Superposition of torquer and suspension fluxes in the ferrite rotor can result in portions of the rotor being pushed far into saturation if an excitation overshoot occurs at the time one torquer is full-on. This phenomenon was demonstrated at both the microsyn and completed instrument levels. The complexities of the IMU circuitry thus far has not permitted the demonstration of this phenomenon at the system level. However, data analysis plus experimental evidence that large scale factor and bias shifts occurred frequently across IMU turn-off and turn-on, gave indications that excitation overshoot effects were a definite possibility.

(3)  Long Term Drift -- Long term curing of the torquer microsyn encapsulating material caused variations in the stresses exerted on the torquer stator by the encapsulant. The permeability of the torquer stator varied as a function of the change in stresses with a resulting change in the scale factor of the instrument. The scale factor change was initially about -150/ppm/month but exponentially decayed to nearly zero in 12 to 18 months. The magnitude of the scale factor change was predictable once a history of scale factor measurements of the unit was established and accounted for.

(4)  Pendulosity Change -- On several early Block II units, the hollow float used in the PIP was not adequately sealed and damping fluid leaked into its interior. The presence of damping fluid in the float produced a variable additive component of pendulosity. Since scale factor is inversely proportional to pendulosity, a large and unstable change in scale factor resulted. The sealing technique used with these units was devised for a less viscous adhesive than was actually used resulting in an improper adhesive flow and a poor bond. An improved sealing technique was developed, however, which corrected this problem.

4.1.3.2 Bias Instabilities. (1) Excitation Overshoot -- As discussed in the previous section, excitation overshoots could have occurred in the Apollo IMU. If the rotor saturation caused by the overshoot affected both torquers equally, only a scale factor shift would have occurred. To the extent that the saturation effects were asymmetrical, a bias shift would result.

(2) Ac Hysteresis -- The term ac hysteresis is applied to bias variations caused by suspension fluxes coupling through fringe portions of the rotor when the float is allowed to rotate to an angular stop. This effect was controlled by adequate specification of the magnetic properties of the rotor material.

(3) Position Memory -- Position memory was an acceleration bias exhibited by the PIPAs. Position memory caused a bias change of 0.1 $cm/sec^2$ for an accelerometer input ranging from 0 to $\pm$ 1 g. The change in position memory bias as a function of imput acceleration was predictable and thus was compensated for since the expected acceleration profile was known. It was thought that eddy currents in the stator were a contributing factor to position memory. The eddy currents caused the flux levels in the stator to be frequency dependent. In the presence of an acceleration, the frequency of the binary torquing (either plus or minus) would vary and, as a result, the average flux levels would likewise vary. The net result was an acceleration dependent imbalance between the plus and minus torquing loops. A possible cure for this phenomenon is currently under investigation and involves using a torque generator stator material of higher resistivity thus reducing the magnitude of the induced eddy currents. Preliminary tests indicated that this approach has some merit.

(4) Reversal of Torque Location -- The original PIP design had the torque generator located at the end of the instrument that contained the mounting arrangement. It was noted that bias changes were affected by stresses applied to the torquer by the clamping forces used during PIP alignment. The torquer was placed on the outboard end of the PIP, and a major source of bias changes was removed.

(5) Fluid Impurities -- Contaminants or gas bubbles within the damping fluid can become positioned between the float and the case of the PIPA exerting error torques on the float. Proper filling techniques would minimize the possibility of foreign particles in the fluid at the time of manufacture. Gas bubbles could result from air leaking into the instrument from outside or from helium gas leaking into the fluid from the hollow float or from the bellows.

A cemented metal band was used to seal the joint between the two end housings of the PIP. This joint, as well as the float and bellows seals, was carefully checked before filling the unit with damping fluid. Filling the instrument was accomplished through small fill holes at either end of the case. Once filled these holes were sealed using metallic compression seals.

By use of the above technique, the possibility of air leaking into the instrument was virtually eliminated. Slow leakage from the float could easily be kept to acceptable levels with known cementing and fabrication techniques.

The bellows remained one of the most critical elements in the design. Rigid screening tests on the bellows and careful inspection by X-ray techniques after completion resulted in bellows failures becoming rare occurrences.

4.1.3.3 Input Axis Instabilities. ( 1 ) Fluid in the Float -- As discussed in the above section on scale factor instabilities, some Block II units leaked damping fluid into the hollow float resulting in pendulosity changes. The pendulosity changes were both in magnitude and direction, the directional changes causing an IA shift within the unit. In the case of these units, the IA shifts obtained were large ( $>10 \, \widehat{\min}$ ).

( 2 ) Fluid Behind Rotors -- During the Block II program, a minor IA shift problem was discovered on some units. The rate of IA shift was in most cases less than 20 $\widehat{sec}$ / day. The cause of the shift was traced to the existence of small crevices behind the rotor which became only partially filled with damping fluid at the time of fill. The fluid within these crevices was then free to move within a limited channel behind the rotor as a function of the position of the instrument relative to the gravity vector. Recent changes in the rotor assembly technique were expected to eliminate this problem.

( 3 ) Suspension Hang-Up -- Suspension hang-up is caused by the air gap between the suspension stator and rotor on a given suspension becoming large enough to cause the unit to operate at, or slightly beyond, resonance. Because of build tolerances which cannot be easily tightened, a small possibility exists that an Apollo pendulum may hang-up if the float is allowed to settle into an axial extreme. As a precautionary measure, all PIPs were required to pass a suspension hang-up test which insured that the float would re-center from any possible rest position.

( 4 ) Temperature Sensor -- After several failures with the original sensor, it became apparent that the sensor leads should have been attached to the flat wire sensor on the PIP with a strain loop. This design change was complied with early in production. A resistance check using a megohmmeter was introduced in order to eliminate any unit which might have suffered a shorting-type failure resulting from assembly pressures.

( 5 ) Moding Change -- Utilization of the binary torquing loop in the Apollo PIPA resulted in a stable oscillation of the float about its null, or minimum TG output signal. The number of torque pulses occurring during each positive or negative going half-cycle was an indication of the operational mode; i.e., two pulses alternately in each direction was defined as a 2:2 mode. The mode actually obtained for a given instrument was dependent on a number of parameters, including the angular moment-of-inertia of the float, the TG sensitivity, the magnitude of the torque pulse, and the viscous damping.

Early Block I units moded a mixed 3:3 plus 4:4. The presence of the 4:4 mode required additional computer time. Accordingly, the viscous damping in the PIPA was increased to a value which would insure the elimination of the 4:4 mode.

(6) Gaussing -- Gaussing is a term used to describe bias changes resulting from variations in the magnetic state of the TG rotor/stator pair. The rotor operates as part of the suspension and torquer magnetic circuits. Current overshoots in either of these two circuits would have produced non-reversible changes in the magnetic operating point of the rotor material. Furthermore, large changes in rotor orientation such as torquing of the rotor to a rotary stop because of a malfunction in the torquing circuit would have caused changes in the magnetic state as a result of disaccommodation.*

In an adequately controlled environment, gaussing would not constitute a problem. However, the Apollo system was not immune to transients (particularly during turn-on, turn-off sequences), and gaussing did occasionally occur.

Theoretically, degaussing is the elimination of a gaussed condition by the application of a high level ac flux to the torquing circuit. The flux level is reduced gradually to zero and in theory should result in no magnetic hysteresis in the material. In practice, however, the material was found to exhibit some residual hysteresis after degaussing. If the degaussing operation was always performed in an identical manner, a fixed residual was obtained, with a corresponding repeatable bias level. However, some variations normally occurred in the manual degauss cycles with resultant bias variations. The use of an automatic degausser optimized the degauss cycle repeatability and normally resulted in a stable reference bias level.

4.1.4 PIP Repair/Redesign Program.

An improved configuration was developed for the PIP repair program. The redesign was instituted in an effort to eliminate failures resulting from bubbles and accumulations of fluid in or around the float. Sample instruments were built and

*Disaccommodation is a time-dependent variation in the permeability of a magnetic material following a change in the magnetic field within the material.

the design was verified. NASA then contracted with Sperry to do failure analysis and verification, rebuilding and testing of 50 PIPs, the majority of which were internal repair units.

4.1.5 Conclusions

With the experience of seven years, numerous conclusions may be drawn regarding the good and bad features of the program.

(1) Since the basic PIP represented a more sophisticated version of the Polaris PIP, which had been in production for some time, most technical problems occurred in external areas of the new instrument such as modules, cabling, connectors and alignment rings.

(2) The importance of a reasonable degree of personnel and organizational stability cannot be overemphasized. MIT maintained a constant group on the task, whereas the Sperry engineering and production groups experienced some shifting of personnel. The program by and large was a successful one, certainly to the extent that changes, improvements, and corrections, where required, were integrated into production with a minimum of impact to an uninterrupted delivery of PIPs to the program.

(3) Monitoring of instrument performance in the field, coupled with an active failure analysis and corrective action program, provided the means for revealing any deficiencies in the design or construction of the instrument. As the quantity of occurrences of each recognized example of degraded performance or field failure reached a statistically meaningful total, analytical and correlative techniques became effective in diagnosing the basic cause of the problem thus permitting appropriate corrective action.

(4) Any reliability test program aimed at determining the life characteristics of a component such as the Apollo accelerometer must be most carefully planned and executed if meaningful results are to be achieved. The 16 PIP life test program was initiated with the purpose of determining an MTBF for this component of the Apollo guidance system.

The planning of this test program proved difficult from the very beginning. First, there were no recognized normal field failure modes for the 16 PIP. The build process and the acceptance test cycle was designed to eliminate the likelihood of electrical shorts or opens and particulate matter in the damping gaps, these being the only recognized failure modes. With only nine units assigned to the program, it was recognized that thousands of hours of operation might well generate no failures at all. This turned out to be the case. Performance limits were therefore set up, more or less arbitrarily, to provide some measure of the operational reliability of the instrument. Some attempts were made to determine environmental effects but, because of the limited number of sample units available, no definitive results could be obtained. Because of the unavailability of a sufficient number of regular test consoles and the desire to avoid impacting the acceptance testing schedule, special, relatively inexpensive, electronic run-in equipment was designed and constructed.

The instruments were operated on a long term, continuous basis on this set-up and transferred to a regular test station at periodic intervals for official data checks. Puzzling performance differences were noted when comparing results obtained at the electronic run-in station versus those secured at the console test station. The inconsistencies in the data could not be explained by the test and engineering information then available. Eventually, the reasons for these inconsistencies were learned, and it was then apparent that the life test setup was not suitable for evaluating the instrument performance. The effort was therefore terminated.

## 4.1.6 Recommendations

The results of seven years of experience with the 16 PIP MOD D, as used in the Apollo guidance system, have shown the instrument to be basically an excellent one for the intended purpose. The effect of system interfaces on performance have been throughly studied. The deficiencies of the instrument have, to a great extent, been revealed and corrective action taken to eliminate or minimize them.

A limited number of conditions still remain for which an explanation and effective corrective action have not yet been developed.

(1) Analysis of repair unit failure data revealed that a sizable percentage of units experienced a shift in uncompensated bias in excess of that which could be ascribed to test equipment variation. Investigation into the cause of this shift should determine whether the shift is stable, thus permitting the use of these instruments and eventually allowing development of corrective action to eliminate this condition.

(2) Certain anomalies in the bias performance characteristics have been identified as resulting from disaccommodation in the ferrite material or the PIP rotor. Work is underway to improve the magnetic structure of the ferrite in order to minimize the disaccommodation.

(3) Certain other anomalies in the bias performance characteristics do not appear to be related to the disaccommodation phenomenon, and the cause is as yet undetermined. This anomaly is identified as position memory.

(4) There is another magnetic effect associated with gaussing which causes both bias and scale factor variations in performance. An effort to make use of an ultra-sensitive airbearing torque-to-balance loop in an attempt to identify and quantify this effect is underway.

## 4.2  Inertial Reference Integrating Gyro (IRIG)

### 4.2.1  General Description

The Apollo Block I IRIG was basically a modification of the type of gyro used in Polaris, that is, a single degree of freedom, floated integrated instrument. Its wheel assembly was supported by a pair of preloaded angular contact ball bearings adjusted to give an isoelastic structure having equal compliance along the spin and output axes. It was enclosed in a sealed, spherical beryllium float. The wheel assembly was driven by a hysteresis synchronous motor in an atmosphere of helium which prevented corrosion of the ferrous parts and also provided good heat transfer within the float. The float assembly was suspended in a brominated fluorocarbon flotation fluid which provided flotation and damping. The fluid was fractionally distilled to yield polymers of approximately the same length and nearly the same viscosity. This minimized fluid stratification under operating and storage conditions. A ducosyn at each end of the unit performed the task of signal and torque generation.

A ducosyn consists of a separate magnetic suspension assembly, a separate transducer microsyn mounted as a single coaxial-coplanar unit, and contains two separate stators mounted to the case and two separate rotors mounted on a common ring on the float assembly. The signal generator ducosyn was mounted on the positive output axis end of the float to provide magnetic suspension and serve as a transducer to provide an electrical analog signal proportional to the position of the float. A torque generator ducosyn was mounted on the negative output axis end of the float to provide magnetic suspension and serve as a transducer to convert an electrical error signal into a torque about the output axis.

The IRIG characteristics are summarized in the following table.

| ELEMENT | TYPE | SPECIFICATIONS |
|---|---|---|
| Wheel assembly | 28 Volt, 2 phase, 800 Hz | Angular momentum: $0.434 \times 10^6$ gm cm$^2$/sec at 24,000 rpm |

| ELEMENT | TYPE | SPECIFICATIONS |
|---|---|---|
| Damper | Brominated Fluorocarbon Density 2.385 gm/cc at $137^{\circ}$ F | Damping coefficients: about OA: $4.75 \times 10^5$ dyne cm/rad/sec about IA: $1.5 \times 10^9$ dyne cm/rad/sec about SRA: $1.5 \times 10^9$ dyne cm/rad/sec |
| Signal generator | 8 pole E-connected micro-syn 4V, 3200 Hz | Angle sensitivity: $10$ mV/mr |
| Torque generator | 12 pole modified E-connection with reset and bias compensation windings. | torque sensitivity $T^{(+)}$ or $T^{(-)}$ to common: $0.6$ dyne cm/$ma^2$ |
| Magnetic suspension | Tapered, $20^{\circ}$ included angle | Radial stiffness 6 gm/0.0001 inch Axial stiffness 0.8 gm/0.0001 inch |
| Temperature Sensors | Two thermistors on each end housing suspended in fluid | Resistance: 345 ohms at $137^{\circ}$ F Resistance gradient: 6 ohms/$^{\circ}$ F/thermistor |
| Mechanical Characteristics Prealignment unit | (1) Shroud jacket: provides magnetic shielding and vacuum envelope for the gyro. (2) Size: 2.810 dia $\times$ 5.10 in. (3) Weight: 1.97 lbs. | |

Figure 4-9 is an illustration showing the main features of the gyro instrument. During final IRIG assembly, a prealignment package was added to the signal generator end of the gyro case. Addition of this package made the gyro a prealigned unit. The package contained the following components:

(1)  Suspension capacitors for ducosyn suspension.

(2)  Temperature sensor normalization.

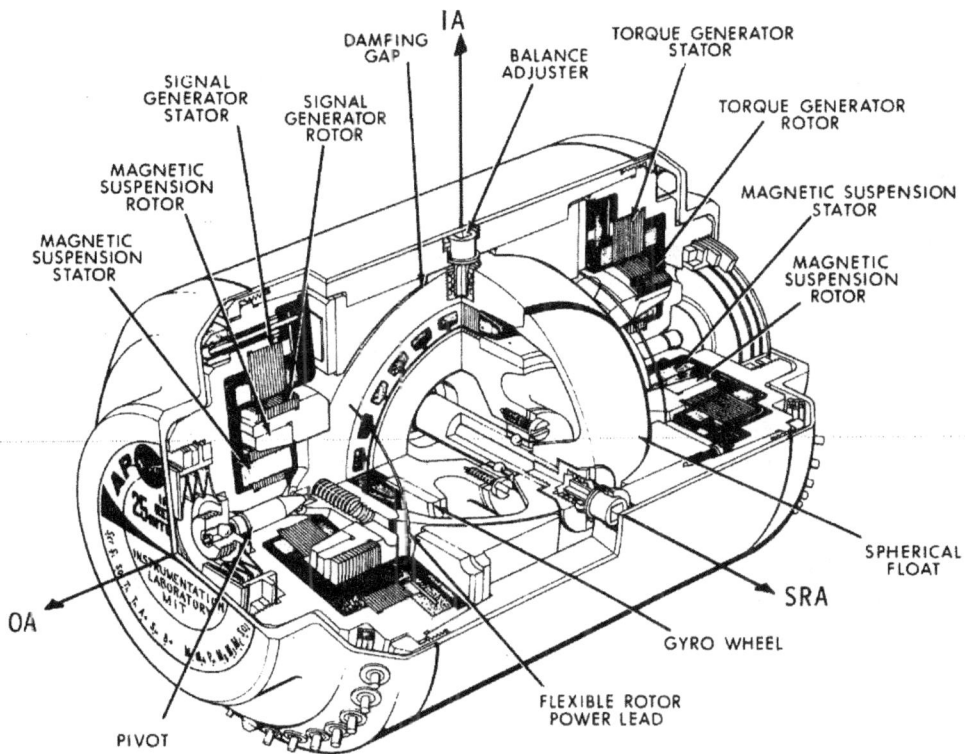

Fig. 4-9 Apollo IRIG

(3) End mount heater prealigned to the gyro input axis.

(4) Torque generator normalization.

(5) Signal generator preamplifier with normalized gain.

The gyro was prealigned on a test stand with the input axis aligned to a slot in the mounting ring. The alignment was carried over to the IMU stable member where a pin was precisely located to pick up the slot. The use of prealigned components made assembly techniques simpler, and brought about excellent correlation between component and system performance of the gyroscope.

The prealignment hardware consisted of a heater and end mount assembly, signal generator preamplifier, and normalization networks (See Figure 4-10). The heater and end mount assembly maintained the gyro at proper operating temperature and performed the task of accurately maintaining input axis alignment with respect to the IMU stable member or test stand. The preamplifier amplified the signal generator output and normalized the signal generator-preamplifier gain product for the instrument.

The normalization networks adjusted the suspension current, torque generator sensitivity, bias compensation current, and temperature sensing resistance. The prealigned gyro constants were:

(1) Suspension current phase : $-45^{\circ} \pm 3^{\circ}$

(2) Temperature indication at $137^{\circ}$ F: 769.6 ohm $\pm 1.0$

(3) Bias drift adjustment: 0 $\pm 5$ meru[*]

(4) Gyro transfer function: $1200 \pm 30$ mV/mr.

(5) Pulse torque scale factor: $\pi/2^{20}$ $\pm 500$ ppm rad/pulse

### 4.2.2 Apollo I Gyro Problems

The Apollo I gyro problems were related for the most part to magnetic suspension, the torque generator, and manufacturing.

(1) The gyro bias drift was different depending on which torquer winding was excited last (positive or negative). This was the result of residual magnetism left in the torquer core after torquing current was removed. To compensate for this drift component, the torquing was programmed to always end a torquing sequence in the same direction, that is, the last pulse was always a $T^{+}$. This left the bias torque resulting from the residual magnetism phenomenon always in the same direction. The magnitude of this term remained constant within reasonable bounds and was compensated for by the Apollo Guidance Computer.

---

[*] 1 meru = 1 milliearth rate unit = 0.015 degree/hr.

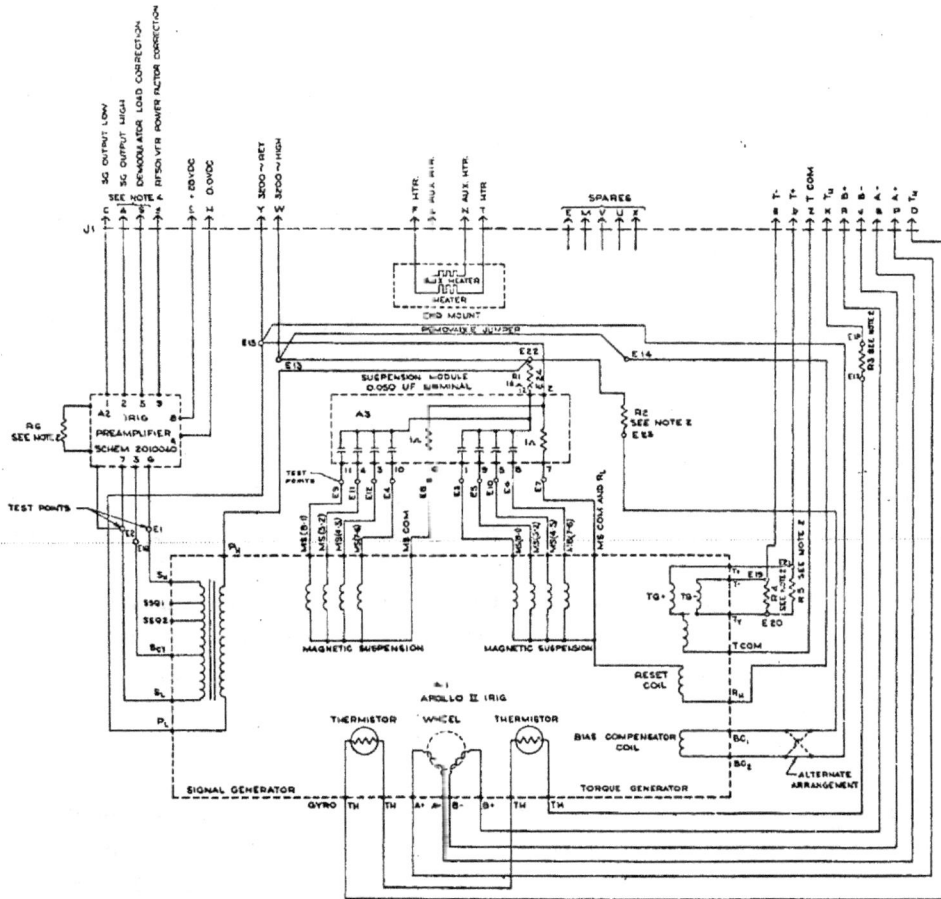

Fig. 4-10 Apollo II IRIG Prealignment Network

(2) The gyro floats displaced radially when the torquer was energized. This displacement resulted from asymmetries that were attributable to manufacturing tolerance buildup, and to tolerances associated with how closely the magnetic suspension centering could be achieved. Some units that were displaced to their jewel extremes during torquing resulted in up to 2% pulse torque scale factor uncertainty (The required stability was 0.1%.) The value of the magnetic suspension damping resistor was correspondingly decreased to stiffen up the magnetic suspension thus reducing the radial displacement when the torquer was energized.

(3) Some of the first units developed acquired suspension shorts to the gyro end housing. These shorts were caused by insufficient insulation between the end housing printed circuit boards and the mu-metal shields. This problem was recognized and corrected by providing more clearance between the printed circuit pins and the mu-metal shield, and by increasing the thickness of insulation between the printed circuit board and the shield.

(4) A design improvement increased the efficiency of the gyro motor by changing the hysteresis ring material from Graphmo to Simonds 73. Because of scheduling problems, this change could not be incorporated into all units and resulted in gyro units utilizing both materials. The system wheel supply compensation had to be selected depending on the distribution of gyros in the system using Graphmo/ Simonds hysteresis ring material. For this reason, a gyro change in an IMU could require a wheel supply compensation change, hence causing a program impact by not supplying true gyro interchangeability.

4.2.3  Block II Changes

4.2.3.1 Reset Winding  During the Block I phase, it became evident that significant improvements could be made in gyro design. Further improvements in the gyro ushered in the second generation or Apollo II instrument. The Apollo II gyro maintained the features of the Apollo I with the exception of ducosyn design. In redesigning the ducosyns, both the signal generator core and torque generator core were changed from 8 poles to 12 poles, and a reset winding and bias winding were added to the torque generator. The introduction of the 12 pole core made possible the incorporation of the reset or wash winding as it is sometimes known. This winding was excited with 3200 Hz, positioned on the poles in such a manner as to provide no net torque to the output axis, and was most effective in washing out the residual magnetism following a torquing pulse command. The reset winding permitted simplification of the torquing logic and eliminated the need for guidance computer programming to account for the bias term ordinarily present. The bias adjust winding made possible the nulling of all non-g sensitive terms within the gyro proper.

4.2.3.2 Mu-Metal Shield. The 12 pole torquer did introduce a torquing rate sensitivity to external magnetic fields that required incorporation of a magnetic shield in the form of a mu-metal shroud end cover. Installation of this shield was required as a retrofit to some of the Apollo II gyros.

4.2.3.3 Configuration Design. The magnetic suspension was redesigned to provide axial as well as radial support for the float while increasing the radial restoring force.

Redesign of the end housing to accommodate the 12 pole ducosyn also included two other significant changes: The end housing was designed as a single piece providing (1) better control on the concentricities of the suspension and microsyn, and (2) micro-inch fit of the pivot, thus eliminating a radial centering problem that had limited the final assembly process yields of the Apollo I gyros.

A comparison of the Apollo I and II gyros is shown below.

| | APOLLO I | APOLLO II |
|---|---|---|
| Wheel Package | ◄———————— Same ————————► | |
| 3200 Hz Excitation | 2 Volts<br>~100 ma | 4 Volts<br>~250 |
| Signal Generator | Single Ended<br>No Quadrature Tap<br>8 Poles | Center Tap Sec<br>Quadrature Taps<br>12 Poles |
| Suspension | No Axial<br>Radial 2 gm/0.0001" | Axial 0.8 gm/0.0001"<br>Radial 6 gm/0.0001" |
| Reset | Two Positive Pulses<br>After Torquing Negative<br>Winding | Reset Winding<br>4V, 3200 Hz, 50 ma<br>Continuously Demag-<br>netizes Torquer |
| Bias Winding | No Bias Adj.<br>Bias < 30 meru | Bias Adjustment Bias<br>< 5 meru Accep. Test<br>< Uncertainty Level |

| | APOLLO I | APOLLO II |
|---|---|---|
| Torque Generator | 8 Pole V<br><br>$+$<br>TM⁺ ⌇⌇ ⌇⌇ TM⁻<br>(SEC)   (SEC) | 12 Pole Y<br><br>$+$   $-$<br>SEC ⌇⌇ ⌇⌇ SEC<br>⌇⌇ PRI |
| Prealignment | $\frac{H}{C}$ Normalized<br><br>Preamp on S.M | Torque Normalized<br><br>$\frac{H}{C}$ K Normalized<br><br>Preamp with Gyro<br>Bias adjusted |

4.2.4 Apollo II Gyro Problems

(1) The first Apollo II instruments showed a torquer sensitivity which was influenced by external magnetic fields. The incorporation of a mu-metal end bell on the torque generator end housing reduced this magnetic field sensitivity to a negligible level.

(2) Prealignment hardware, initially, had a high incidence of solder joint failures. The problem was caused by stresses in the solder joints resulting from flexure of a printed circuit board as it was mounted in the prealignment case. Consequently, the printed circuit board was made thicker, and the method of mounting was changed to minimize flexure. These measures effectively eliminated the problem.

(3) The gyro preamplifiers were susceptible to high frequency noise. A capacitor placed at the preamplifier input eliminated this problem.

(4) A large number of float freedom failures, resulting from particle contamination within the units, appeared in the early phases of Apollo II. Improved control over the cleanliness of the assembly area, tighter control over assembly procedures and a general improvement in workmanship and quality control made a significant improvement.

(5) Approximately 50% of early gyro failures were the result of bearing failures (<1000 hours). This failure rate caused such a shortage of gyros that the

use of workhorse or nonflight status instruments to complete system checkout and selloff were required. A repair program featuring a new bearing design was introduced. However, as a result of startup delays and somewhat modified manufacturing procedures, the output of repaired gyros was too low to significantly reduce the severe gyro shortage.

The reason for bearing failures continues to be under investigation. To date, however, the principal reason for an increased failure rate relative to the Block I bearing has not been discovered. The wheel package was unchanged from that of the Apollo I gyros which exhibited a much better bearing life history. Various cleaning procedures, lapping methods, and retainer redesigns were attempted without conclusive results (See Section 4.2.7 for a more detailed presentation).

### 4.2.5 Failure Prediction

4.2.5.1 Reliability Test Program. At the start of Apollo system design, a gyro reliability test program was initiated. Two aims of the test program were: (1) to obtain a failure model for predicting impending system failures, and (2) to determine whether intermittent operation, as planned for the Apollo guidance system to save power, was any less desirable than continuous operation. A total of 10 gyros was divided into two groups: six were operated continuously for 200 hours followed by servo performance for 24 hours. Four were operated intermittently, i.e., with wheels operating for 3 hours and off for 12 hours for a total of 210 hours followed by servo performance for 24 hours. Except for servo runs, all running hours were accumulated on a torque-to-balance loop. More than 52,000 wheel hours were accumulated on the 10 gyro units. Four wheel bearing failures occurred during the accumulated running time.

Figure 4-11 compares the acceleration sensitive drift about the input axis (ADIA) data for units that were run continuously with units that were run intermittently. For the same operating time, the intermittently operated units resulted in the more stable performance. Therefore, the intermittent mode of operation in the system did not result in a performance penalty. There were no differences in the BD and ADSRA terms for the two modes of operation.

4.2.5.2 Data Flow. Data on all the significant parameters of gyro performance were collected for all units from the first component tests through all ISS and G & N tests. These data were evaluated and formed the basis for failure prediction. Moreover, procedure was established for centralized assessment of all gyro and

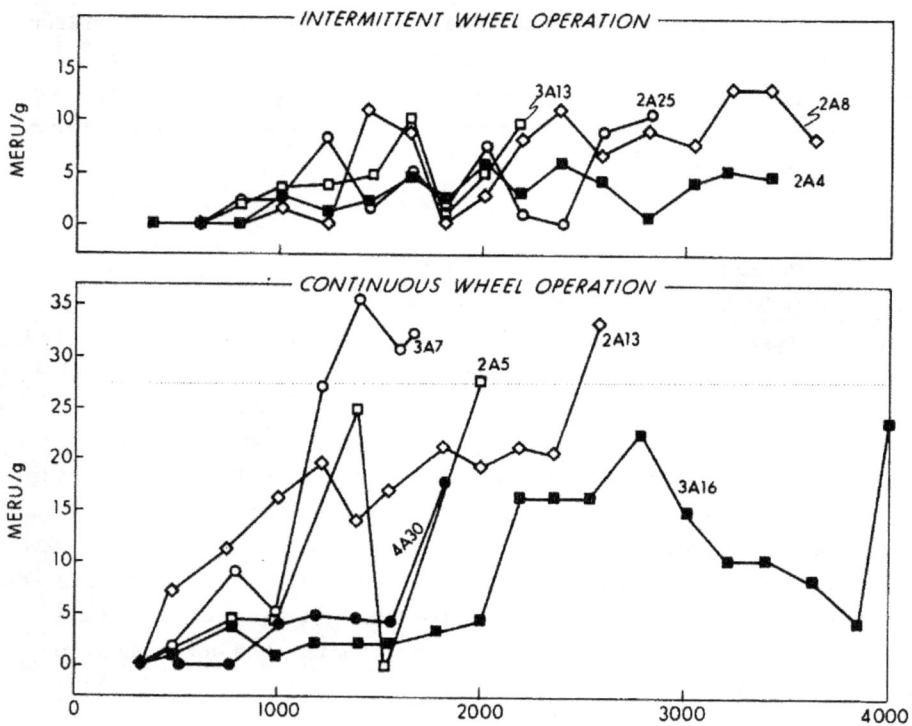

Fig. 4-11  Wheel Operating Hours

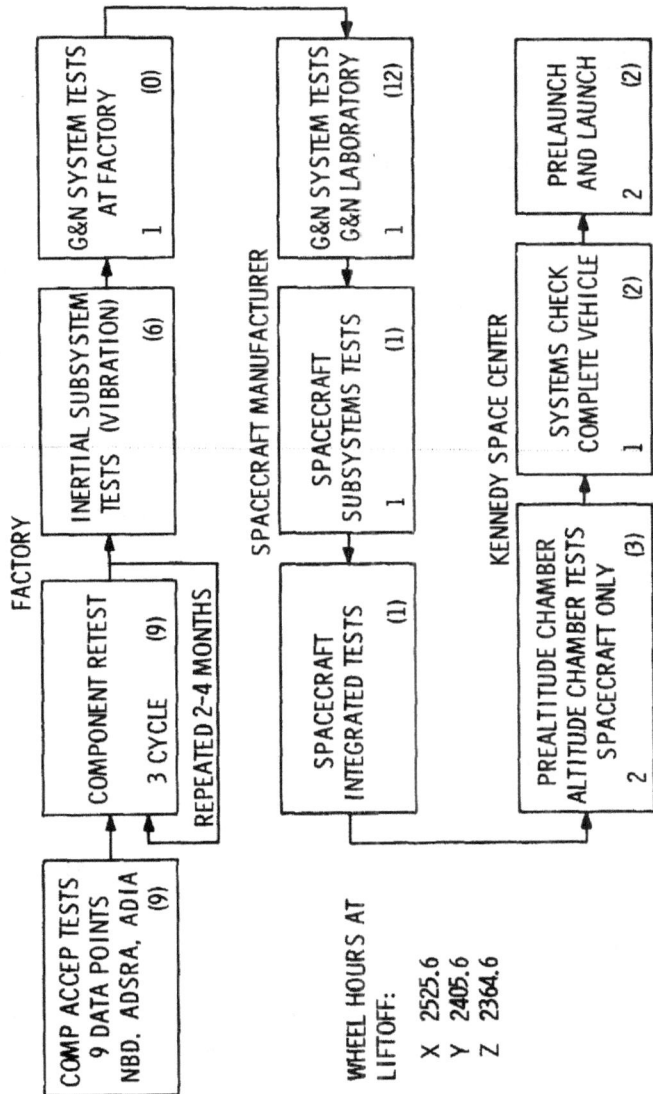

FACTORY

| COMP ACCEP TESTS 9 DATA POINTS NBD. ADSRA, ADIA (9) | → | COMPONENT RETEST 3 CYCLE (9) REPEATED 2-4 MONTHS | → | INERTIAL SUBSYSTEM TESTS (VIBRATION) (6) | → | G&N SYSTEM TESTS AT FACTORY 1 (0) |

SPACECRAFT MANUFACTURER

| SPACECRAFT INTEGRATED TESTS (1) | → | SPACECRAFT SUBSYSTEMS TESTS 1 (1) | → | G&N SYSTEM TESTS G&N LABORATORY 1 (12) |

KENNEDY SPACE CENTER

| PREALTITUDE CHAMBER ALTITUDE CHAMBER TESTS SPACECRAFT ONLY 2 (3) | → | SYSTEMS CHECK COMPLETE VEHICLE 1 (2) | → | PRELAUNCH AND LAUNCH 2 (2) |

WHEEL HOURS AT LIFTOFF:

X   2525.6
Y   2405.6
Z   2364.6

S/C 011 Z GYRO DATA TAKEN INDICATED BY NUMBERS IN PARENTHESIS.   OTHER NUMBERS REPRESENT NORMAL PERFORMANCE EVALUATION DATA POINTS.

Fig. 4-12   Inertial Component Performance Data Sequence

accelerometer data. The checkout procedures required that a copy of each data record be sent to MIT. The flow diagram for data from a single gyro acceptance test through flight is shown in Figure 4-12. The collected data were converted to punch cards and magnetic tape in formats suitable for electronic data processing and printing. Additionally, a monthly listing was printed both as an inventory and as data tabulation.

4.2.5.3 <u>Gyro Acceptance Plan</u>. The gyro acceptance plan was based on the design-reference lunar landing mission (DRM) of 200 hours total flight time. Of this time, the gyro wheel was operated 20 hours including the operational time required to initialize the GN & C system and to perform an inflight or ground alignment of the inertial measurement unit (IMU) and, also, operation during thrusting periods. The maximum cumulative wheel operating time, including the final ground checkout operation and a margin of safety, approached 300 hours. This value was therefore chosen as an upper limit on the time of operation for maintenance of angular momentum.

Another constraint on gyro acceptance was the performance necessary to insure a successful mission. Figure 4-13 shows the probabilities that gyro performance changes would result in marginal mission performance. For example, changes in IA acceleration sensitive drift of less than 1000 meru/g among any of the three gyros would not create errors equal to those listed in Figure 4-13, thus 1000 meru/g was the upper boundary for change in gyro performance. Unless drift rate curves for a flight gyro lie to the left of the appropriate curve, a marginal mission is indicated.

Inflight measurement of gyro performance, to correlate gyro performance with mission performance, was considered, but measurement of the acceleration sensitive drift along the input axis (ADIA) and acceleration sensitive drift along the spin reference axis (ADSRA) appeared impractical. One possible method would have been to make a position and velocity error comparison between ground and onboard state vectors, and then to have inferred the specific gyro terms that caused the deviation. The bias drift (BD) terms could have been measured in free fall. Then, after successive IMU alignment, the changes in angular errors measured during a time interval would yield BD. Two important aspects of the gyro acceptance plan included drift performance and system tests, and performance indicators.

(1) Drift Performance Tests & System Tests -- Drift performance data were generally obtained utilizing an inertially stabilized loop (See Figure 4-14). The gyro instrument was mounted on a test table such that the input axis was parallel

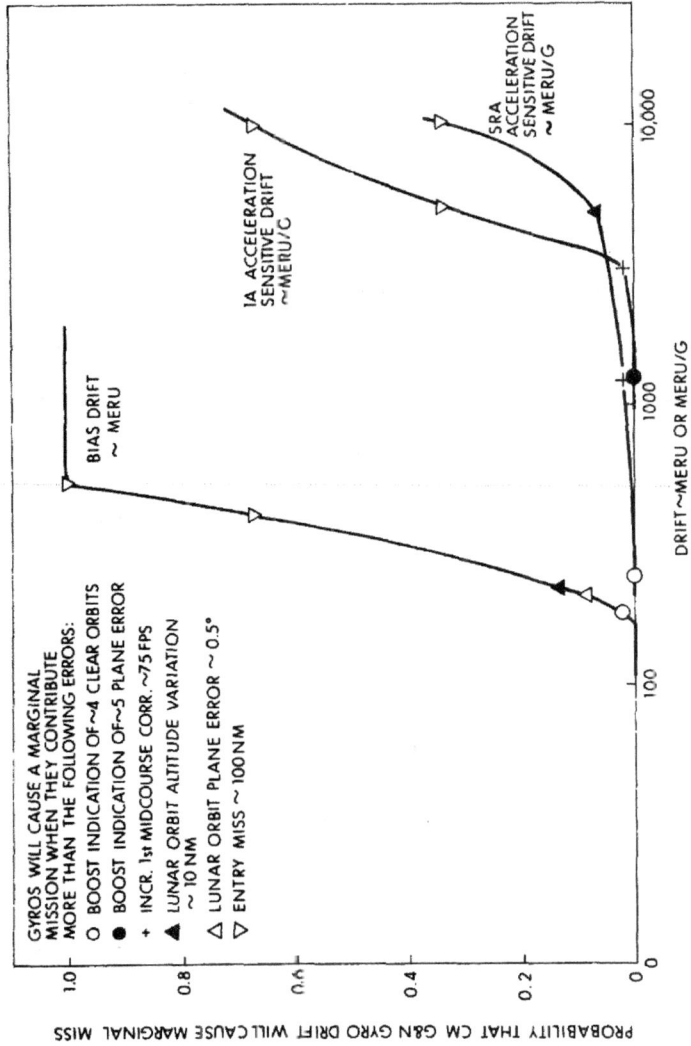

Fig. 4-13 Marginal Mission

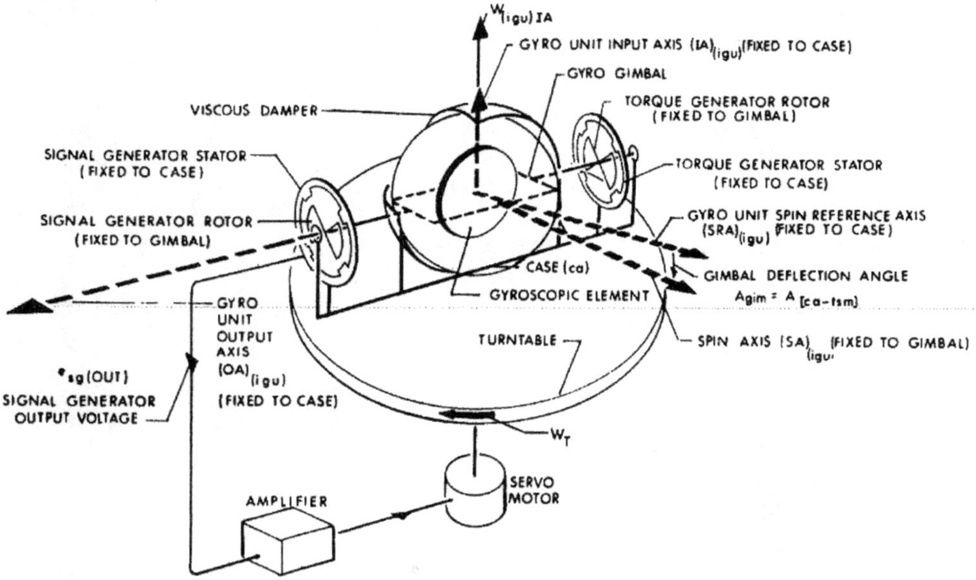

Fig. 4-14 Line Schematic Diagram of the Apollo IRIG Operating
in a Single-Axis Orientation Control System

to the table axis. The gyro signal generator output was amplified and fed to a table servo loop. The output of the table servo loop drove the table torque motor and rotated the gyro about its input axis. In this loop, an ideal gyro would have maintained its input axis fixed in inertial space. As viewed from the earth, the test table would have rotated by an amount equal to the component of earth rate sensed along the gyro input axis. Drifts (deviations from earth rate measured in meru) would occur as a result of torques that act about the gyro output axis. Since the magnitudes of these drift rates can be compensated, the drift rate stabilities provided a measure of the performance of the gyro instrument. A table rate, determined from the time required for the table to rotate through a prescribed angle (e.g., $1^O$), gave the average drift rate over a period of time. By consecutively positioning the table axis so that gravity acted along three gyro axes, the ADSRA, ADIA, and BD drift terms could be determined by solving simultaneous equations. The stabilities of these drift terms were determined by repeated measurements after periods of storage time or by a continuous measurement of table drift rate for a fixed orientation with respect to earth rate.

The design and operational goal for mean time between failures (MTBF) was 100,000 hours. That is, for a group of 500 gyros that passed the performance-prediction criterion during preflight operations (with a probability of 67 percent), not more than one failure was expected among the 500 during the ensuing 200 hours of wheel operation of each gyro. The performance tests measured the torques on the float and the float pendulosity about the output axis. Other tests measured parameters that were performance indicators, e.g., the retainer beat frequency and the stability of the required driving power.

(2) Performance Indicators -- The rotating ball-bearing retainers produced a sinusoidal torque about the gyro output axis at a frequency corresponding to the rotational velocity of the retainer, but since the average value of this sinusoidal torque was zero, there was no resulting drift error. The torque was detectable at the gyro signal output as a sideband of the signal generator frequency (3200 HZ) and the retainer frequency (243 Hz). The magnitude of this signal varied periodically as one retainer lapped or passed the other retainer. A change in the lap frequency was caused by a change in speed of one of the retainers. This frequency stability was found to be related to the ADIA stability. Figure 4-15 shows data for ADIA deviation versus retainer beat frequency for 25 gyro units. The data for each unit were obtained from 12 hour inertially stabilized drift runs with the gyro input axis vertical. Data points for most units fell close to a linear relationship line. This technique was developed during the Apollo gyro production program. Evaluation of this technique indicated significant advantages over many methods now in use; it's simple and easily implemented.

The stability of the power required to operate the gyro motor indicated its condition relative to such variables as frictional torques and gyro wheel bearings. A good power trace has excursions of 5 mW, or about 0.1% of total wheel power. Figure 4-16 shows traces taken at two times during the life of a gyro. At time A, the power excursions were low, but at time B, a significant decrease appeared in the stability of bearing torques. Finally, the wheel failed to reach synchronous speed and bearing failure occurred. The milliwattmeter was used both to screen bearings before they were installed in gyros and to determine whether there was bearing deterioration on completed instruments. Power monitoring was performed at component-level testing and on gyros installed in the IMU.

System tests of gyro performance measured gimbal angles or changes in angle. One method used was to orient the gyro input axis in a direction as to have zero drift, then, finding the gyro orientation with respect to the earth, the drift performance was calculated. A second method was to measure an angular velocity by the use of the gimbal angle transducers in the IMU. The base was assumed to be non-rotating with respect to the earth.

### 4.2.6 Failure Criteria

The failure prediction technique had to have a high probability of rejecting bad gyros and still had to be reasonably easy to implement. If at all possible, the technique was to use parameters that also had a direct bearing on mission performance.

4.2.6.1 <u>Delta-25 Criterion.</u> From a study of the long term drift characteristics of the gyro population, it was found that the ADIA term for 90 percent of the population was stable to better than 25 meru/g. Since wheel bearing deterioration was the principal failure mode for this instrument, a failure criterion related to a 25-meru/g shift in the ADIA term appeared to offer a promising prediction device. This criterion (Delta 25) required that the 25-meru/g shift occur at a single test location to avoid shifts caused by test equipment variability, that is be verified by a second data point to eliminate bad data, and that it occur across a storage period of less than four months because shifts during long storage were noted in this instrument.

Figure 4-17 is a plot for unit, serial no. 3A16. The bearings of this particular unit failed in the reliability program. At approximately 2100 wheel hours, the ADIA changed by more than 25 meru/g. The gyro was in the same location, and the performance change was verified by a second test. At this point, it was predicted that this gyro would fail. For the next 1500 hours, its performance was very respectable in that the total acceleration-sensitive drift along the input axis was less than

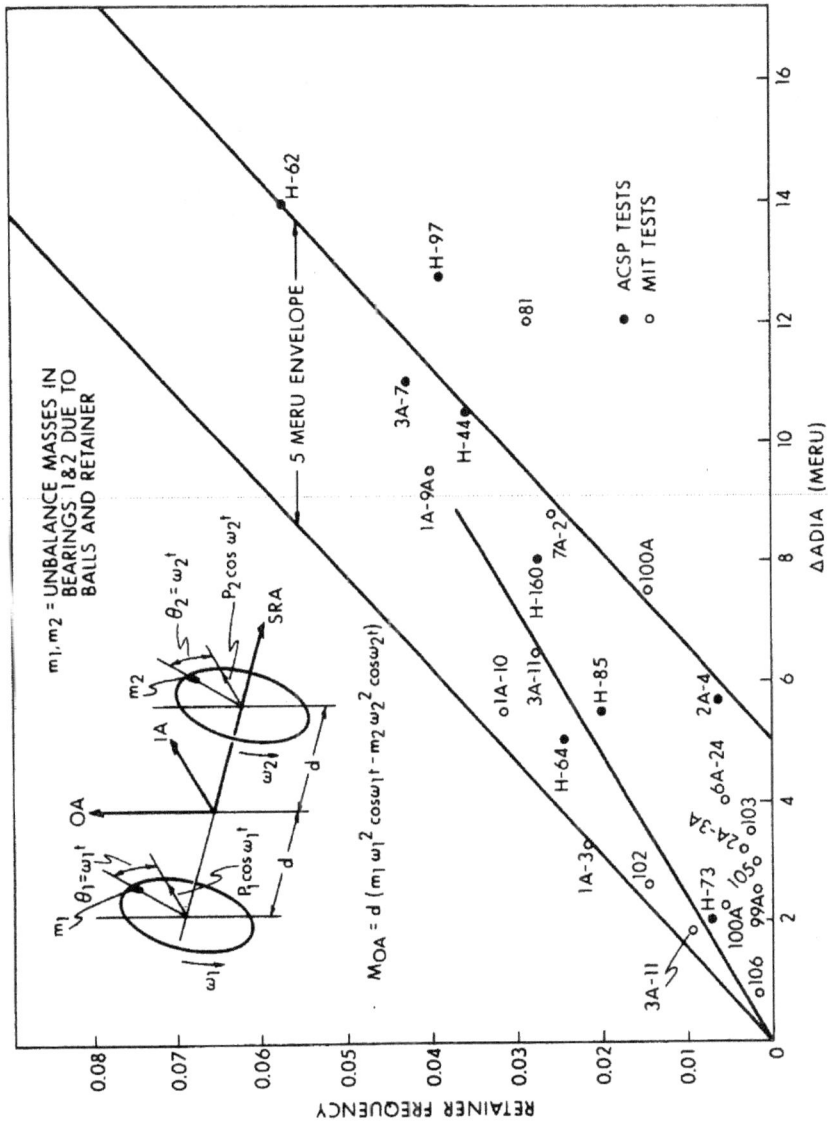

Fig. 4-15  Change in Retainer Lap Frequency vs ADIA for 25 Units

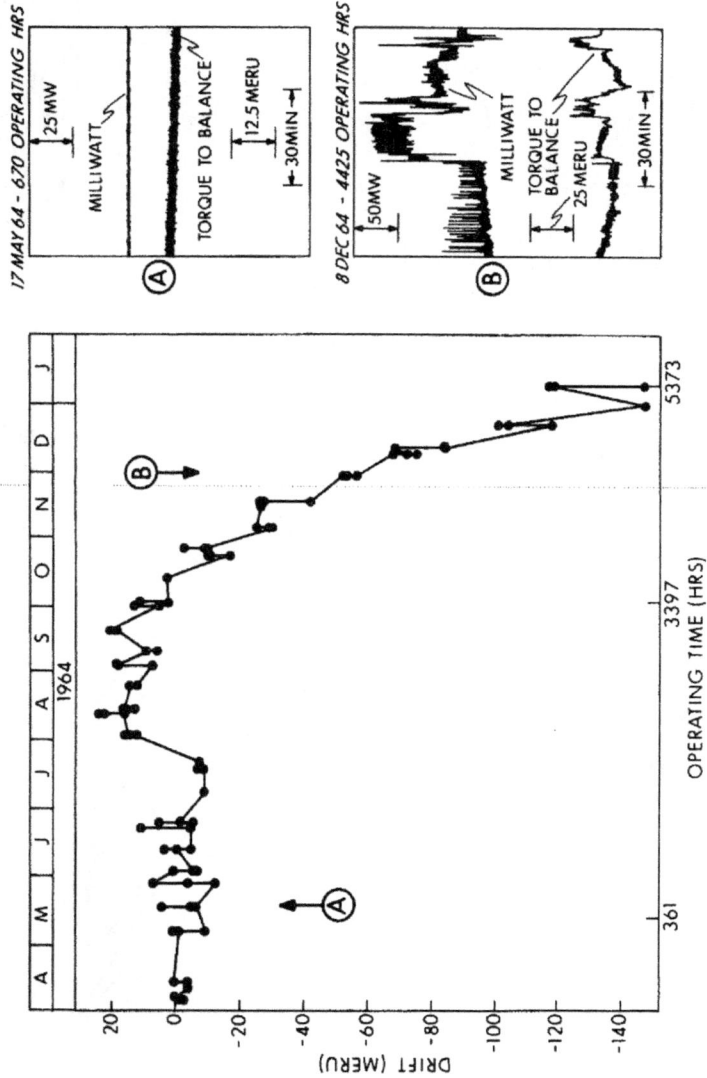

Fig. 4-16 ADIA Drift and Wattmeter Sample Traces vs Operating Time

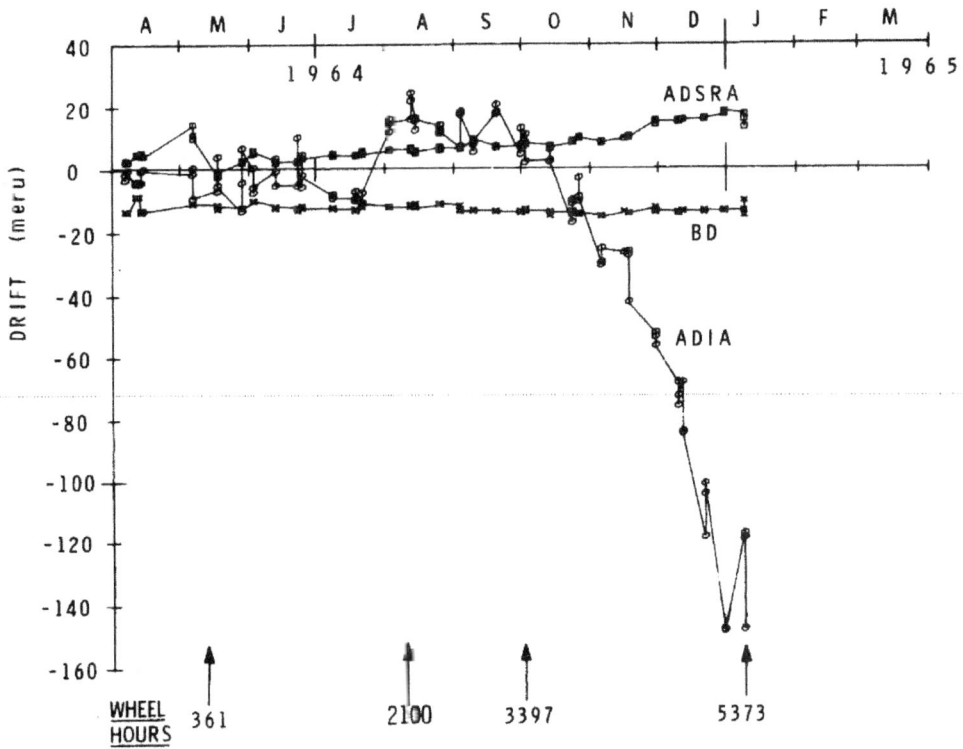

Fig. 4-17  25 IRIG-3A-16 Drift Coefficients

30 meru/g. At about 3600 hours, the ADIA changed by more than 25 meru/g, and after another 1800 wheel hours the gyro failed completely. After each Delta-25 indication, the total change in ADIA for the ensuing 300 hours was less than 28 meru/g, providing a long lead time before the unit failed. During this lead time, gyro performance was still suitable for an Apollo mission. Note that the bias drift (BD) was very stable, and the acceleration-sensitive drift along the spin-reference axis (ADSRA) changed very slowly during the life of this gyro. Small or gradual changes in BD and ADSRA were characteristic for all wheels that failed.

Figure 4-18 is an example of prediction of system performance for a gyro installed in the guidance system. The ADSRA term changed by more than 25 meru/g at the time of installation into the system. Subsequent tests verified this change. The Delta-25 criterion would not have picked up this parameter on that instrument since it changed location. After installation in the system, a subsequent ADIA change of more than 25 meru/g occurred and was verified. Upon this indication of a potential failure, the gyro was removed from the guidance system and was tested until the wheel failed. The wheel time accumulated after the Delta-25 criterion was equivalent to 80 lunar missions. Again, the BD and ADSRA terms remained stable until the gyro failed completely.

The Delta-25 criterion is now used for all gyros. To implement its use fully, the performance of each gyro was measured periodically. Units that exceeded the criterion were either removed as failures from the system to be verified or were requalified. The requalification test was a rerun of the unit acceptance test.

4.2.6.2 The F Criterion. Since there was a correlation between ADIA instability and impending wheel failure, any parameter that measured ADIA stability could have provided a basis for failure prediction. There were three criteria that could be applied: (1) a magnitude change in ADIA exceeding some level, e.g., the Delta-25 criterion; (2) the standard deviation of the ADIA exceeding some level; and (3) the exceeding of some level by the rms value of successive ADIA differences (rmssd). All of these methods were found to obtain large magnitudes for a situation that occurred frequently but was not an incipient failure. For example, a change in drift could have occurred from dimensional changes in the float structure, with the drift then becoming stable at a new level. It was difficult to find a standard magnitude-or-deviation criterion that would account for a stable level change. The F criterion, however, which was based on method 3, served this purpose. By the F criterion, a failure was predicted if the rmssd of the last 11 ADIA data points exceeded 10 meru/g. A unit requalified if its rmssd later dropped below 10 meru/g. The requalification indicated a unit that shifted in ADIA drift but was later stabilized. The rmssd was defined by the following equation:

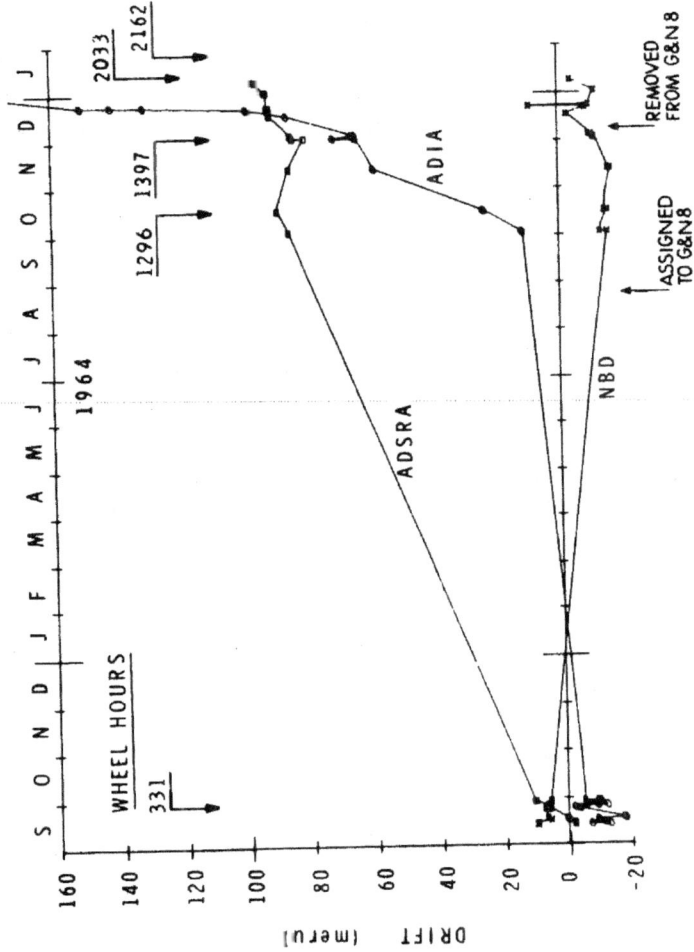

Fig. 4-18  25 IRIG 2A-11 Drift Coefficient

$$rmssd = \left[ \frac{1}{N} \sum_{i=1}^{N} (x_i - x_{i-1})^2 \right]^{1/2}$$

where $x_i$ was the present value of ADIA, and N was 11 for the F criterion.

Figure 4-19 shows rmssd versus wheel hours for unit 2A5, which exceeded the F criterion at 1500 wheel hours. 500 hours later it failed to start at the application of wheel voltage. These hours would have been sufficient time to complete two preflight tests and missions. Gyro 2A11 (Fig. 4-20) exceeded the F criterion during acceptance testing (300 wheel hours), later was requalified, then exceeded the F criterion again at 1200 hours. It had stable ADSRA and BD performance even after it had exceeded the F criterion for ADIA data. It was run intermittently for another 800 hours before failure; again, there was adequate warning to have allowed a successful completion of the mission.

4.2.6.3 Usefulness and Efficiency Indices for the Criterion. Usefulness and efficiency indices are defined in Figure 4-21. The usefulness of a failure-prediction method was considered as the percentage of gyros that were predicted good and, in fact, remained good. The efficiency of the prediction technique rested in the percentage of gyros not rejected. For example, if 5 percent of the good gyros had been rejected, prediction efficiency would have been 95 percent. For the Delta-25 criterion, the calculated usefulness indices for the Apollo Block I and Block II gyros were 99 and 98 percent, respectively. The efficiency indices were 89 and 98 percent.

For the F criterion, all but one of the Apollo Block I bearing failures were predicted. However, 23 units were predicted to fail but did not. This yielded a usefulness index of 99 percent and an efficiency index of 80 percent. For Block II, the F criterion's usefulness index was 99 percent and its efficiency index was 96 percent.

4.2.6.4 Failure Criteria Summary. Failures were grouped into major categories and were as follows:

(1) Bearing Failures -- This group contained non-start wheel units with ADIA instability, decreasing wheel run-down time, and erratic wattmeter data indicating an impending wheel failure. The group also included units that failed from other causes, such as contamination, that is, when failure analysis showed the wheel bearings to be badly deteriorated and imminently susceptible to failure from this cause.

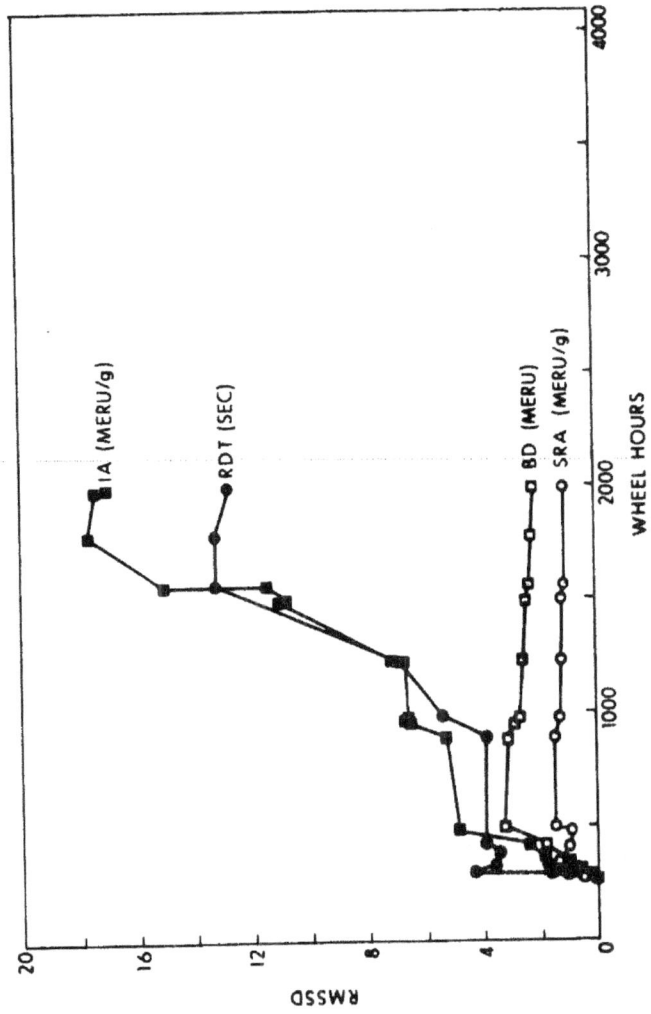

Fig. 4-19 RMSSD vs Wheel Hour – Apollo Gyro 2A5

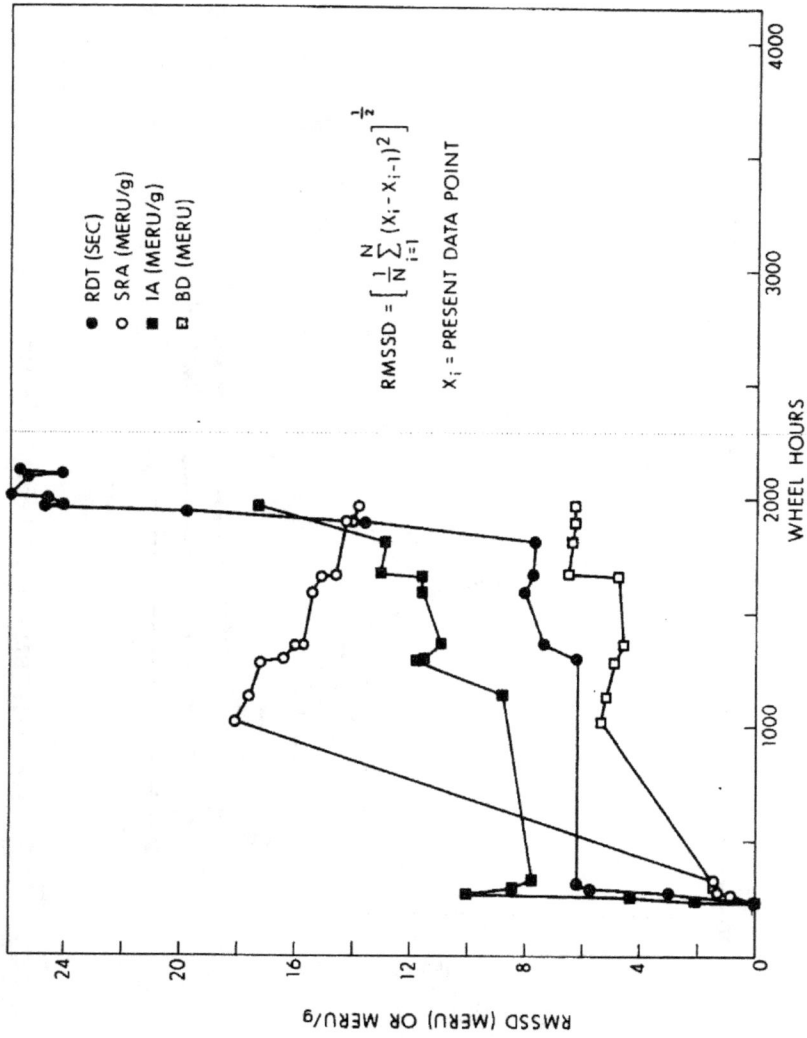

Fig. 4-20 RMSSD vs Wheel Hours - Apollo Gyro 2A11

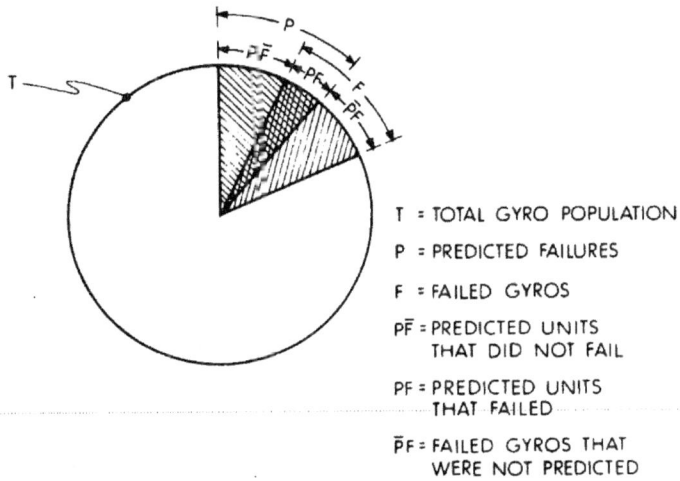

T = TOTAL GYRO POPULATION

P = PREDICTED FAILURES

F = FAILED GYROS

$P\bar{F}$ = PREDICTED UNITS THAT DID NOT FAIL

PF = PREDICTED UNITS THAT FAILED

$\bar{P}F$ = FAILED GYROS THAT WERE NOT PREDICTED

$$\text{USEFULLNESS INDEX} = \frac{T - PF - \bar{P}F}{T - PF} \times 100$$

$$\text{EFFICIENCY INDEX} = \frac{T - P\bar{F} - PF}{T - PF} \times 100$$

Fig. 4-21  Usefulness and Efficiency

(2) Float-Freedom Failures -- This group contained instruments failing the float-freedom test. This test identified units containing foreign fluids or particulate matter in the damping fluid. Most units, which had been identified for the repair program as Category III repairs, evidenced this problem. Other items in Category III were gyro flexlead and ducosyns, i.e., problems associated with the internal gyro but not with the bearings and internal float assembly.

(3) Flex-lead Failures on Centrifuge Testing -- This group consisted of two units that had failed during centrifuge testing at NASA/MSC. Failure analysis had indicated an open wheel lead. Because of this environment testing, these units were not candidates for repair.

(4) Electrical Failures -- This group included typical shorts or opens in the gyro external electrical circuitry and prealignment components. It represented all units grouped in the repair program as Category II repairs.

(5) Delta-25 Predicted Failures -- This group consisted of units predicted to fail by the Delta-25 failure criterion, i.e., a gyro exhibiting a shift in drift terms of more than 25 meru/g was suspect and subjected to further testing and analysis before being committed to flight. The criterion was qualified in that only shifts of 25 meru/g or more occurring at a single test location were considered. Shifts had to be verified by a second datum point, and must have occurred across a storage period of less than four months. The single test location minimized any data shifts caused by test equipment variability. The verification by a second datum point was necessary to eliminate the likelihood of bad data. The restriction of the storage period between tests limited the storage-sensitive shifts that were noted in this instrument.

4.2.7  Failure & Reliability Summary

The Apollo I gyros were subdivided during manufacture into five groups according to the date of manufacture. Most of the discrepancies in drawings and procedures were corrected during the construction of the first two groups. The failure rate for each failure cause was plotted for each group (see Figure 4-22). Induced failures were not included as failures in this plot. The large number of failures, 73 percent in Group I, was attributed to the changes required and to start-up problems. Groups II to IV show a continuing improvement in the percentage of failures, a low of 13 percent for Group IV. Figure 4-23, a plot of the average wheel hours for each group, shows that the average operating hours for the last four groups had decreased slightly. This could partly explain the decreasing number of failures for Groups III and IV and should have resulted in fewer failures for Group V.

ELEC = ELECTRICAL
NOT SPEC = NOT SPEC AT ACCEPTANCE
FF = FLOAT FREEDOM

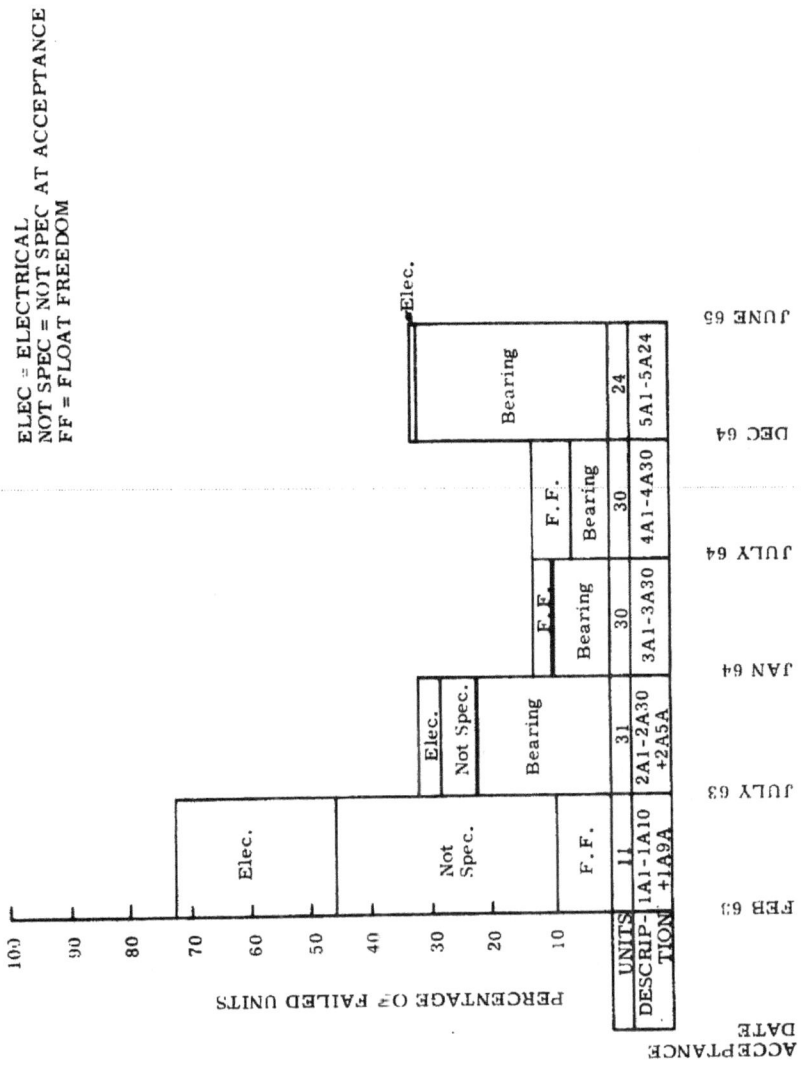

Fig. 4-22 Apollo I Gyro Failure

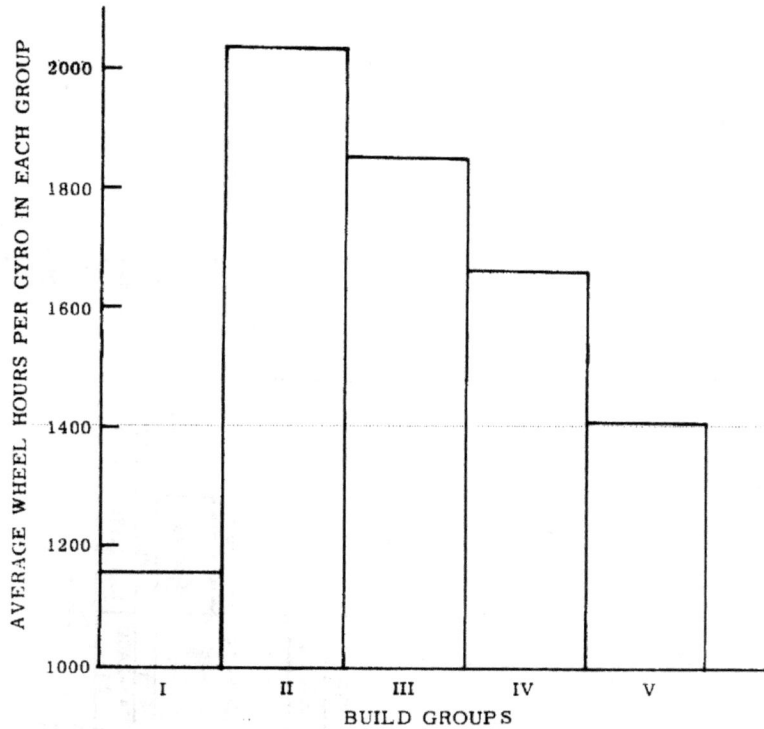

Fig. 4-23  Average Wheel Hours of Apollo I IRIGs by Build Groups

However, the percentage of failures for Group V made a drastic jump to 33 percent. No evident explanation was available to account for this trend reversal. Since the increase was almost entirely the result of an increase in the bearing failures, it was likely that the problem was a wheel-build deterioration, either in the choice of bearings or in the performance of the wheel-build procedures.

A similar study of Apollo II assembly results was made and plotted as shown in Figure 4-24. The groups were determined by dividing the total Block II build program into eight groups of about 30 units per group according to their acceptance date. The results compared to the Block I experience were startling. The failure rate in the first group was 60 percent and succeeding groups varied from that rate by only 20 percent. Bearing failures were a consistent percentage, 25 percent of each total; and float-freedom failures, which were almost negligible in Block I, were a steady contributing factor at about a 20 percent failure rate. Even the first group, which was expected to be of lower quality because of production start up effects, was typical of the entire production run.

Again, the results in the last group were atypical. The float-freedom failures rose to 40 percent, double the incidence in the earlier groups, while bearing failures were drastically lower. The lower bearing failure rate for this last group could have been caused by a lower average operating time for that group. As in the Apollo I analysis, the average wheel hours per group were plotted. The result, shown in Figure 4-25, indicates that the wheel hours for Group 8 were lower than for the other groups. If this was the cause of the improvement in Group 8, bearing failure rather than operating hours, could result in the last group being the worst of Apollo II gyros.

From an analysis of the failures, several interesting observations were made. Failures were uniformly distributed over the range of time from acceptance and were therefore not time-dependent. The majority of the failures occurred in the first 1200 wheel operating hours, but no particular failure mode was completely screened out at this early interval. Any unit that was removed from a system for a problem but not verified as a failure was classified a good unit for this analysis.

An increasing failure rate trend starting at 2100 total wheel hours was observed. The small population of units in these life intervals (less than 20 units) provided a low confidence level. Moreover, it was not possible to define a "wear out" point with the data available. Therefore, the following conclusions were applicable:

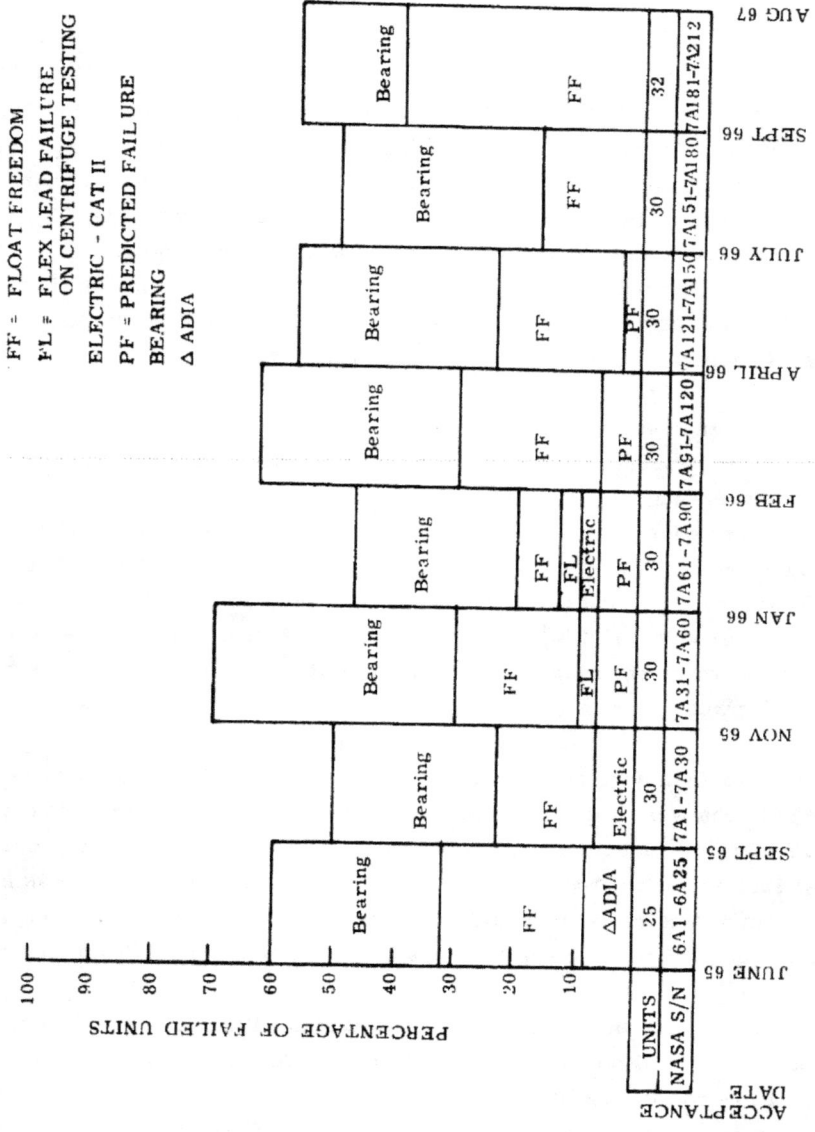

Fig. 4-24 Apollo II Gyro Failure and Predicted Failures

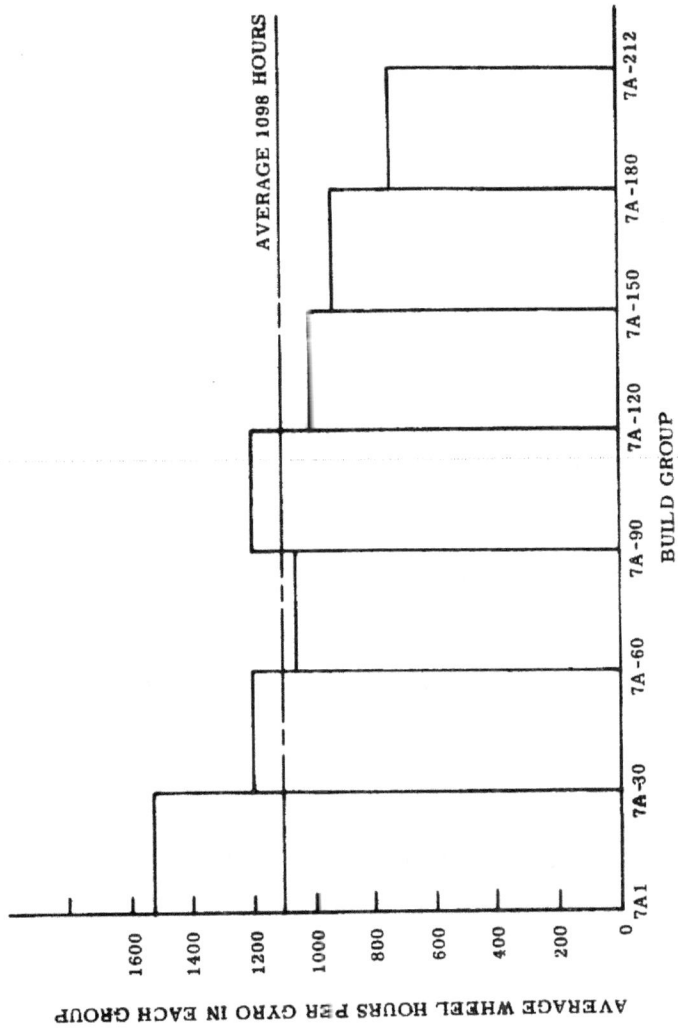

Fig. 4-25 Average Wheel Hours of Apollo II IRIGs by Build Group

(1) The failure rate per wheel life was consistent at 20 percent per 300 wheel hours and at 14 percent per 5 months. The difference in these two failure rates indicated that the gyro population was not being subjected to an average accumulation of 300 hours over five month intervals.

(2) Sixty-nine percent of the failures were identified in the first 1200 wheel hours after assembly. This was the result of classifying all Apollo II units soon after acceptance as either potential failures or flight-quality units. For these units, a longer acceptance cycle (1200 wheel hours) would have been very effective in screening a majority of the failures.

(3) It was estimated that 14 failures would have occurred if all the Apollo II gyros were operated an additional 300 hours, and 12 of those failures would have occurred in the ensuing five months.

(4) Inasmuch as catastrophic failures resulting from poor solder connections or contaminant in the damping fluid could not have been predicted, such failure modes were eliminated or held very low.

Although the remaining failure mechanisms for the Apollo gyro were quite complex, the reliability study program yielded two useful criteria for the primary failure mode or wheel failure. The Delta-25 failure criterion was not 100 percent effective because it predicted failure for some units that were not about to fail and, conversely, it missed some that did fail. Further, it required elimination of data that occurred across location changes, even though data were shown to correlate very well at all locations. As more data were accumulated, more refined prediction criteria were established. The F criterion was one such promising refinement. Indices both of its usefulness and efficiency were high for the Apollo Block II gyros, and it was easier to apply than was the Delta-25 criterion, since the F criterion could be applied to all test data regardless of storage period or of test location changes. The parameter monitored both for the Delta-25 and the F criteria was ADIA. Neither BD nor ADSRA showed significant changes that could be correlated with wheel failure.

As of January 1, 1967, 361 production gyros had run a total of 380,771 wheel hours. Excluding the wheel hours on any units after failures were predicted by the Delta-25 criterion, the total wheel operating hours were 346,175, which yielded an MTBF of 100,000 hours with 70 percent confidence.

Another analysis of failure data related bearing-failure rate as a function of wheel hours for Block I and Block II. This analysis was aimed at determining if a wear-out time was evidenced for the gyro bearing. A rise in the Apollo II bearing-failure rate at about 2100 wheel hours seems to have been indicated. However, the population was too small, that is, the confidence level was low as a result of the large effect of a single failure in these groups to have allowed this conclusion. As the population operating hours increased, however, a more definitive conclusion became possible.

One very interesting finding was that the average bearing failure rate for Apollo II was 10.4 percent for 300 wheel hours, which was three times the bearing-failure rate for Apollo I (3.1 percent). The cause of the higher bearing failure rate for the Apollo II gyro could have only been caused by a change in assembly, material lots or processing, or quality control monitoring between these two gyros, since there was no design change to the wheel package between these two instrument builds. By every failure criteria presented, Apollo I gyros operated within acceptable performance requirements far longer than Apollo II instruments. The difference in life characteristics between the two groups had not changed substantially with usage and the passage of time.

There were some assembly and test changes that occurred between the Apollo I and Apollo II build effort. It was not possible to tell whether any of these contributed to the poorer reliability obtained for Apollo II units. However, some of these differences are included here for clarification.

(1) The bearings were directly assembled into the float for most of the Apollo II build, whereas the Apollo I bearings were first tested in a dynamometer and then disassembled, cleaned, inspected, and assembled into the float.

(2) The GSE torque-to-balance loop and ramp generator were changed between the Apollo I and Apollo II build. The quantity of ramps required for Apollo II float freedom test was decreased from that required for Apollo I.

(3) There was a significant change in the reliability documentation required for Block II. This configuration required documentation and traceability for each component in the build. The large documentation requirement may have created a false sense of security and resulted in a reduced monitoring effort in the important areas of assembly and test.

( 4 ) The Apollo I gyro contract changed from a cost-plus, fixed-fee basis in the middle of its build to a fixed-price plus incentive for the remainder of the Apollo I build. The Apollo II contract was a cost-plus incentive contract.

( 5 ) Apollo I was built while the Polaris production contract was in-house at ACE, using the same assembly personnel and area. The higher level of available Navy quality control monitoring may have helped Apollo reliability.

( 6 ) As a result of phasing out of Polaris and other programs at ACE, there was a reassignment of the assembly and test technicians working on the Apollo program. A higher level of personnel reassignments appeared to have occurred during the entire Apollo II program which resulted in a continuous retraining effort.

The large Apollo II failure rate strongly points out the inadequacy of the acceptance test as a fully effective reliability screen. A more meaningful series of specifications for assembly, inspection and test to achieve an effective acceptance screening, appears to have been required.


4.3    Power and Servo Assembly (PSA)

4.3.1  Description

The Power and Servo Assembly ( PSA ) comprised the electronics that supported the operation of the inertial and optical subsystems and was located in a common module for convenience in packaging and spacecraft installation.

4.3.1.1 Block I PSA. The Block I PSA contained most of the electronics associated with the inertial subsystem, including the IMU servo electronics, IRIG and PIP pulse torque electronics, CDU electronics, power supplies, and mode switching electronics. Some of the ISS electronics were located on the IMU and included IRIG and PIP microsyn suspension networks, signal generator preamplifiers, temperature control sensors, and resolver adjustment networks. Phase correction capacitors were located on the CDUs. The Block I PSA also contained most of the electronics for the optical subsystem.

The design of the Block I PSA was influenced largely by the concept of in-flight malfunction diagnosis and repair. Repair was to be done by direct replacement of failed subassemblies with on-board spares. The Block I PSA was designed with this concept as a ground rule. To this extent, the PSA electronics were packaged in small, replaceable modules that plugged into a wired tray. Where possible, functional subassemblies having similar requirements were packaged as identical modules. Thus, all three IMU gimbal servo amplifiers were packaged as identical modules, and each contained three values of a feedback element appropriate to the gain requirements of the inner, middle, and outer gimbal serovs.

Selection of the proper feedback was effected by tray wiring when the servo amplifiers were plugged into the inner, middle, or outer gimbal positions in the tray. Thus, one spare gimbal servo amplifier could function as the repair module for any of the three locations. Similarly, identical 1% and 5% 800 Hz power supplies were used for the inertial and optical subsystems.

The modules were plugged into ten individual trays and appropriately interconnected by tray wiring. The trays, in turn, were plugged into a tray connector and interconnect harness assembly for further power and signal interfacing with the rest of the G& N system and the spacecraft. To aid in inflight malfunction diagnosis, a connector at the front of each tray provided selected test points that were then available to the inflight monitoring test set. The inflight maintenance concept was abandoned after the PSA design was released for manufacture and subsequently was never used.

The electrical problems experienced in the Block I PSA fell into two general categories: those identified from engineering evaluation and qualification testing and reliability analysis, and those identified from System Testing. The schedule for delivery of hardware had not permitted the luxury of an exhaustive evaluation phase prior to manufacturing release. Consequently, designs were being released for manufacture concurrent with breadboard and engineering evaluation testing. As circuit parameters became better defined, component values were altered to place them closer to design centers. Reliability analysis of component failures during the engineering evaluation and parts qualification phases required replacement of some components with more reliable ones. These problems were not considered inherent design deficiencies but reflected a normal refinement of the designs with the passage of time.

System testing exposed design deficiencies that required PSA modifications. They are listed below:

(1) A filtering or decoupling capacitor was added to the ADA preamp to prevent the 102.4 Kpps signal of the fail circuit from being coupled into the servo loop.

(2) A parallel diode circuit was added from the summing junction of the gimbal servo amplifier to complementary side of the input differential amplifier in order to reduce the gain for large saturable signals at the summing junction. This modification resulted because a large charge was trapped on a capacitor in the ADA filter when moding between coarse align and fine align took place.

(3) Capacitors were added to the output stage of the CDU motor drive amplifier to form a high-frequency break in the response of the amplifier. These capacitors corrected a problem that had resulted because of variations in the gain-bandwidth product of the output transistors.

(4) The bandwidth of the demodulator filter in the encoder electronics was reduced by changing the value of the filter capacitors. This was considered necessary to reduce the 6400 pps noise that was being generated by the IMU heater controls and coupled into the encoder. The bandwidth was reduced to approximately 3400 Hz.

(5) The ternary mode of PIP pulse torquing resulted in intolerable accelerometer bias instabilities. The decision to change to binary torquing required a redesign of much of the PIP electronics.

4.3.1.2 Block II and LM PSA. The designs of both PSAs were essentially the same. Since the LM did not have an optical subsystem that required electronics within the PSA, the LM PSA was physically smaller. The basic ISS portions of the CM and LM PSAs were identical, however. The inflight maintenance concept was dropped for the Block II CM and LM. This permitted a more efficient packaging scheme and allowed for hermetic sealing of the PSA. There were, however, certain calibration modules that were matched to a particular IMU. It was necessary to open up the PSA and change these modules in the event that an IMU was replaced, unfortunately with the consequence of compromising the integrity of the PSA hermetic seal. The decision, therefore, was to incorporate portions of the IRIG and PIP pulse torque electronics into separate packages. For the LM, the Pulse Torquing Assembly (PTA), as the package was called, was located close to the IMU and external to the cabin, which was physically separated from the PSA by long cable runs. This location minimized problems associated with long leads. For the CM, the PIPA Electronics Assembly (PEA) was located in the space formerly occupied by the Block I CDUs. The PEA and the PTA were functionally parts of the PSA, but were packaged separately for operational convenience.

The Block II and LM PSA electronics were basically the same as Block I. Changes that were made were to accommodate wider variations in spacecraft supplied prime power voltage, or were refinements of the circuitry based on experience or the result of improved components that became available.

Block I experience showed that the added margin in IMU stablization stability provided by the IMU Angular Differentiating Accelerometers (ADA) was not required. The ADAs and associated electronics were deleted from the Block II design. The CDU was changed to an all electronic device and packaged separately, thus removing the CDU electronics from the PSA. The IMU temperature control was changed from a sophisticated proportional temperature control scheme employing magnetic amplifiers and pulse-width modulation techniques to a much simpler thermostatically controlled on-off type. Thus, all the temperature control electronics from the PSA were eliminated. As with Block I, functional subassemblies were packaged as small modules that plugged into a prewired PSA header. Where possible, functionally similar subassemblies were again packaged as identical modules. The final Block II modules evolved satisfactorily after several component value changes brought about circuitry optimization. The Block II PSA encountered few component-associated problems during the manufacturing period.

Two capacitor failures did occur. One capacitor failed "open" because of inadequate metallization on the end of the roll. Since only one part out of 8000 purchased exhibited this failure mode, it was concluded that no part problem was evidenced by this one failure. Another capacitor "open" was found to be an inadequate internal lead solder connection. The vendor instituted corrective action by using a number of loops for the solder termination to replace the single loop of the original design. A resistor failure was attributed to a fracture of the glass substrate on which the resistance element was formed. After considering the number of devices used versus the one failure encountered, it was concluded that no part problem existed with this type of component.

Two instances of parts-oriented transistor failures were encountered. First, a gimbal servo exhibited a high offset null during module testing. The cause was found to be unstable, long term drift characteristics of the transistor P/N 1010252-1. The corrective action was to add a high temperature bias screen test requirement to the Specification Control Drawing. This screening enhanced the detection of surface contamination of the part. Two instances of cracked dies were discovered in transistors of the 800 Hz amplifier. The transistors were changed to another type; no other corrective action was taken. A design oriented diode problem occurred when diodes P/N 1010385 were found to exhibit a lowering of forward breakdown voltage. This anomaly occurred as the parts were procured later in the program. This condition is generally recognized as an improvement of diode characteristics, but not when used in a voltage-dependent bias circuit such as was the case. A change in resistor value was incorporated in the design to compensate for the change in breakdown voltage.

Changes and modifications to the PSA electronics resulted because of system test results, the inability to procure specified parts over the life of the contract, and the variation of component parameters for large numbers of components. The gimbal servo amplifier required a change in output transistors as a result of discontinuance of manufacture of the originally specified part. Feedback diodes of the servo amplifier had to be tightly specified as the program developed because of parts parameter variations. A design change was incorporated in the 3200 Hz, 1% supply to prevent overshoot of the supply output at turn-on. System tests had shown the overshoot to be a factor in gaussing of the PIP.

### 4.4    Inertial Measurement Unit (IMU)

### 4.4.1  Mechanical Design

The IMU consisted of three single degree-of-freedom gyros (IRIGs) and three single degree-of-freedom accelerometers (PIPAs) mounted on a stable member which was isolated from vehicle orientation by a servo-driven three degree-of-freedom gimbal system. The IMU gimbal system consisted of an outer gimbal mounted to the case, a middle gimbal mounted to the outer gimbal, and the stable member, which served as the inner gimbal, mounted to the middle gimbal. The gimbal arrangement was compatible with the spacecraft's Flight Director Attitude Indicators (FDAI) and had the following defined axes:

(1) The outer axis was fixed to the vehicle and was parallel to, and in the same direction as, the vehicle X-axis.

(2) The middle axis was parallel to, and in the same direction as, the vehicle Z-axis when the outer gimbal angle was zero.

(3) The inner axis was parallel to, and in the same direction as, the vehicle Y-axis when the outer and middle gimbal angles were zero.

The IMU was secured to a navigation base and aligned to the Optical Unit Assembly (OUA) axes within 0.1 milliradian. Figure 3-1 schematically depicts the general gimbal configuration and axis definitions.

The following is a description of the IMU components starting with the stable member and proceeding outwards to the case (Refer to Figure 3-2).

4.4.1.1 Stable Member Components. The stable member was a sintered block of beryllium, machined to accurately locate the inertial sensors and other components. The inertial components were prealigned and electrically normalized on a test stand before they were assembled into the stable member. This saved considerable alignment adjustment at later stages of testing.

The components on the stable member were:

(1) Three size 25 single degree-of-freedom Inertial Reference Integrating Gyro (IRIG) units. These were prealigned with respect to slots in the mounting hardware. They were accurately positioned by means of locating pins precisely located on the stable member. The gyro end-mount supported the necessary electronics resulting in component interchangeability. The electronics included: microsyn suspension networks, temperature sensor normalization networks, temperature control heaters, torque generator normalization networks, and a signal generator preamplifier with normalized gain.

(2) Three size 16 single degree-of-freedom Pulsed Integrating Pendulums (PIPAs). They were also prealigned with respect to the mounting hardware and accurately positioned on the stable member by means of locating pins. Each PIP was provided with an end-mount heater.

(3) Gimbal-mounted electronics which were necessary to service suspension and output signals from the pendulous accelerometers.

(4) Temperature control heaters and sensors which maintained the inertial components at proper temperatures.

(5) A prewired harness-assembly which interconnected the stable member electrical components and also mated with the slip rings on the inter-gimbal assemblies.

4.4.1.2 Components on the Inner Axis. The stable member was supported on the middle gimbal by a pair of preassembled inter-gimbal assemblies, one on each end of the axis. Figure 4-26 depicts typical inter-gimbal assemblies.

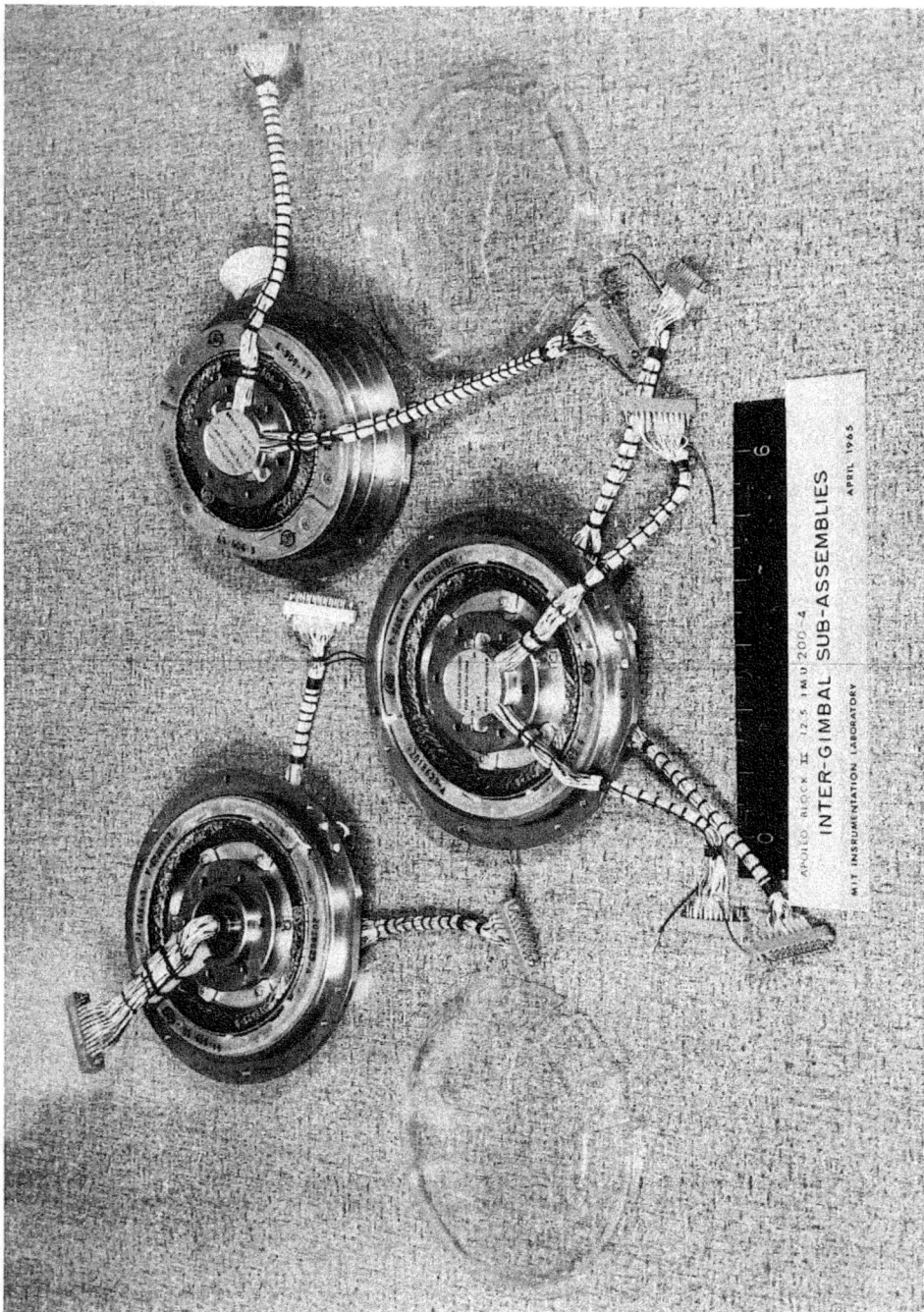

Fig. 4-26 Inter-Gimbal Sub-Assemblies

One inter-gimbal assembly contained a pancake-type dc servo-torque motor
having a 3.8 inch diameter air gap (1.12 ft.-lbs./amp), a 40 circuit slip ring assembly, and a duplex pair of ball bearings preloaded to provide a predetermined
axial yield rate of 7 micro-inches per pound. The inter-gimbal assembly housing
and stub shaft were made of beryllium. The stub shaft supported the torque motor
rotor and slip ring assembly. The slip rings were terminated in connectors which
interfaced with the stable member harness and the middle axis slip rings. The
slip ring wires were clamped to the housing a few inches away from the slip ring
assembly so that they flexed with slight oscillations of the gimbal without the slip
rings and brushes moving in relation to one other. This eliminated slip ring wear
under most operating conditions. The bearings were floated axially to allow for
manufacturing tolerances and thermal expansion.

The other inter-gimbal assembly contained a combination 1 speed-16 speed
resolver transmitter, a 1 speed gyro error signal resolver, a 40 circuit slip ring
assembly, and a duplex pair of ball bearings preloaded the same as those on the
other end of the axis. The stub shaft supported the resolver rotors and slip ring
assembly. The resolvers were accurately aligned, by the resolver manufacturer,
with respect to alignment slots on the stub shaft and housing. These slots engaged
pins on the stable member and middle gimbal. The bearings of this assembly were
fixed and therefore provided axial support for the stable member. The axis of rotation, as determined by the bearings, defined the inner axis, which had unlimited
rotation.

4.4.1.3 Components on the Middle Gimbal. There were no electrical components on
the middle gimbal. The middle gimbal was made up of two hydroformed 0.040 inch
thick aluminum hemispheres pinned and bolted together at brazed aluminum flanges.
There were four additional flanges to which were attached the inner and middle axis
inter-gimbal assemblies. Pins on these flanges picked up the alignment slots of
the resolver inter-gimbal assemblies.

4.4.1.4 Components on the Middle Axis. The middle gimbal was supported in the
outer gimbal by a pair of pre-assembled inter-gimbal assemblies, one on each end
of the middle axis. They were similar to the inner axis inter-gimbal assemblies.
One assembly contained a pancake type dc servo torque motor having a 3.8 diameter
air gap, a 40 circuit slip ring assembly, and a preloaded duplex pair of bearings.
The bearings were floated axially. The other inter-gimbal assembly contained a
combination 1 speed and 16 speed resolver, a 40 circuit slip ring assembly, and a
fixed (not floating) duplex pair of preloaded ball bearings. The axis of rotation as
determined by the bearings defined the middle gimbal axis. The middle gimbal had
unlimited rotational freedom about this axis.

4.4.1.5 <u>Components on the Outer Gimbal.</u>  The outer gimbal was made up of two hydroformed 0.040 inch thick aluminum hemispheres pinned and bolted together at brazed aluminum flanges.  There were four additional flanges to which were attached the middle and outer inter-gimbal assemblies.  Pins on these flanges picked up the alignment slots of the resolver inter-gimbal assemblies.  The outer gimbal had slots in it to permit forced convection of heated air between the middle gimbal and the case.  This was provided by two axial-flow blowers mounted on the outer gimbal.  The slots also permitted assembly of the outer gimbal since the flanges were internal.

4.4.1.6 <u>Components on the Outer Axis.</u>  The outer gimbal was supported in the IMU case by a pair of outer axis inter-gimbal assemblies, one at each end.  They were also similar to the inner axis inter-gimbal assemblies.  One assembly contained a pancake-type dc servo-torque motor having a 3.125 inch diameter air gap, a 50 circuit slip ring assembly, and a preloaded axially floating duplex pair of ball bearings.  The other inter-gimbal assembly contained a combination 1 speed and 16 speed resolver, a 50 circuit slip ring assembly and a fixed duplex pair of preloaded ball bearings.  The axis of rotation determined by the bearings defined the outer axis.  The outer gimbal had unlimited rotational freedom about this axis.

4.4.1.7 <u>Components on the IMU Case.</u>  The IMU case was made from 0.060 inch thick aluminum hydroformed spherical sections with integral roll-bonded, inflated, coolant passages.  Two Seaton-Wilson hydraulic quick-disconnect fittings were attached for connection to the spacecraft water-glycol coolant system.

Mounting pads for mounting the IMU to the navigation base were accurately machined to locate and align the IMU in two axes.  The third axis was defined by dowel bushings which engaged accurate holes on slots in the IMU case and navigation base.

Two 61 pin electrical connectors interfaced with the G & N harness; they were keyed to prevent incorrect mating.  An electrical sixteen speed resolver adjustment module and a blower control relay were located inside one of the end covers that sealed each end of the outer axis.

The case covers were sealed by the use of  o-rings to effect a hermetic seal of the IMU.  This insured that the thermal transport medium (air) within the IMU would not leak out during vacuum operation.  Another benefit of hermetic sealing was that a greater choice of engineering materials would be available inasmuch as the constraint on the use of toxic or odorous materials and lubricants, and the seizing of bearings in vacuum would be minimized.

4.4.1.8 Size and Weight. The IMU case diameter was 12.5 inches; it was 14 inches across the end covers on the outer axis. The volume of the IMU was approximately 1100 cubic inches, and it weighed about 42.5 pounds.

4.4.2 Thermal Design

4.4.2.1 General. The IMU temperature control systems were designed to maintain the gyroscope and accelerometer operating temperature within narrow limits. Thermally, the IMU could have been viewed as three concentric aluminum spheres, the inner sphere as the middle gimbal, the next sphere as the outer gimbal and the external one as the case.

A liquid-to-air heat exchanger covered part of the surface area of the case. Between the middle gimbal and case was an air circuit that used two blowers mounted on the outer gimbal to force air from the area of the middle gimbal towards the case heat exchanger. The IMU heat transfer is shown in Figure 4-27.

The Block I and Block II IMU's were thermally similar. The respective temperature control schemes were also very similar. In both instances temperature deviations from a fixed absolute temperature were detected. These temperature deviations were used to control heater power to the stable member. The implementation of the control schemes was, however, different.

4.4.2.2 Block I Thermal Design. The Block I IMU temperature control design was hampered from the start by an inadequate definition of the environment within which the IMU was expected to perform. In particular, the spacecraft thermal environment, primary coolant loop characteristics, and prime power voltage excursions were unknown. For these reasons, an attempt was made to include much flexibility into the design in order to be able to adjust to the actual environments as they became better defined.

The use of thermal heat of fusion materials to serve as a heat reservoir was considered early in the design. This approach was taken to conserve electrical power. A thermal study, one of the first to define system operation for a lunar landing mission, showed that though this concept was sound, the use of these materials was unworkable based on IMU time line usage. This approach was abandoned and an electronic temperature control system was designed.

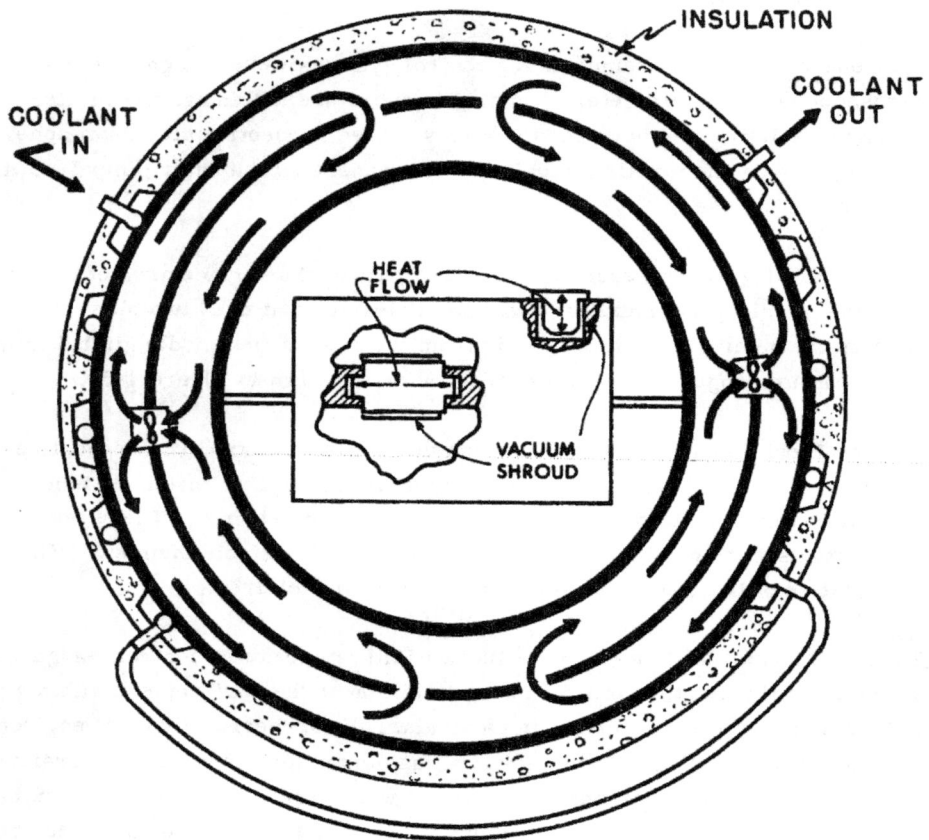

Fig. 4-27 IMU Heat Transfer Diagram

The temperature control scheme utilized resistance wire temperature sensing elements located in the IRIG end mounts. These sensing elements connected in series, measured the average temperature of the three gyroscopes. They formed one arm of a four arm resistance bridge. The remainder of the bridge was located in the PSA. The bridge error signal, proportional to the temperature difference between the actual average gyroscope temperature and the desired temperature, controlled the operation of magnetic amplifiers in the power servo assembly (PSA). These, in turn, provided power proportional to temperature deviation. The power was in the form of a 20V, 3200 pps pulse-width modulated square-wave voltage to the stable member heaters.

An additional set of heaters, controlled by a thermostat on the stable member and powered directly from spacecraft prime power to the G & N system, comprised a redundant temperature control system. This did not provide the precise control of the primary system but was adequate to satisfy the crew safety and mission success requirements.

Temperature sensing thermistors within the gyroscopes were used to monitor the gyroscope temperature. They were connected in series and formed one arm of a four arm resistance bridge. The other elements of the bridge were located in the PSA. The error signal output of this bridge controlled another magnetic amplifier which turned on an alarm light if the gyro temperature exceeded specified limits. It also provided an output for telemetry of IRIG temperature and an output to the front of the PSA tray for use by the in-flight failure monitor.

The temperature sensing resistance elements of the accelerometers were used to monitor PIP temperature in a manner similar to the gyroscope monitoring scheme.

The two blowers on the middle gimbal were used to vary the thermal resistance between the inner gimbal and the case. Saturable reactors on the outer gimbal varied the blower speed as a function of stable member heater power.

4.4.2.3 Block II Thermal Design. The Block II IMU temperature control design effort was undertaken to develop a smaller, lighter, simpler, more reliable temperature control system for the Block II IMU. Advantage was taken of the knowledge and experience gained from the Block I IMU. The Block II spacecraft was essentially identical to the Block I vehicle; therefore, the IMU environment was well known. A good thermal model of the IMU was developed from Block I experience. The uncontrolled IMU heat sources, e.g., inertial components, torquers, resolvers, and gimbal mounted electronics, were well defined.

Temperature control of the Block II was accomplished by using a mercury thermostat as the temperature sensing element in bistable temperature control system. Additional mercury thermostats were used to provide an out-of-limits temperature alarm indication and for control of the two blowers. A sensor of this type was used in the Block I IMU for emergency control. It proved to be accurate, stable, extremely reliable and had a very small dead band. Figure 4-28 is an electrical schematic diagram of the Block II temperature control system.

As with Block I, heaters were located on each end of the gyroscopes and on one end of each accelerometer. Two additional cartridge heaters were embedded within the stable member. Each of the four thermostat modules contained one control, or anticipatory heater, and one bias heater. Padding resistors were placed in series with the heaters to adjust the power in each heater. The temperature difference between the IRIGs and the PIPs was adjusted by properly proportioning the amount of power in each heater.

The temperature control power required to keep the control thermostat within its limit cycle operating range varied in response to environmental and power changes within the IMU. The effect of control power variations on inertial component temperature was minimized by adjusting appropriate PIP control heater and control thermostat anticipatory heater padding resistors. Fixed bias heat was applied to the PIPs to bring them to proper operating temperature whenever the ISS was in the operate mode. The stable member temperature was affected by fluctuations in gyro wheel power. As wheel power increased, for example, because of an increase in wheel voltage, the control power and therefore the stable member temperature decreased. To keep the PIP temperature from decreasing along with the stable member, it was necessary to add more heat to the PIPs. This was accomplished by energizing the PIP fixed bias heaters with gyro wheel supply voltage.

The performance characteristics of the temperature control system are shown in Fig. 4-29. The thermostat cycle of operation at 50% duty cycle was 0.3 while at the same time the gyroscope limit cycle was attenuated to 0.00065 degree F and that of the accelerometers to 0.009 degree F. . The stable member temperature continued to drop as the control power was increased, as one might expect, but the inertial component temperature remained constant throughout the power range. The blower extended the dynamic range of operation, was infrequently used and was limit-cycled by a thermostat. There were two separate sensors to detect temperature extremes and alert the astronauts. In addition, high limit mechanical thermostats were used in every heater power line to prevent an over heated condition. These were set to open the heater power at a temperature of about $10^{o}$ F over the normal control temperature. These were rarely used but when necessary prevented damage to valuable equipment.

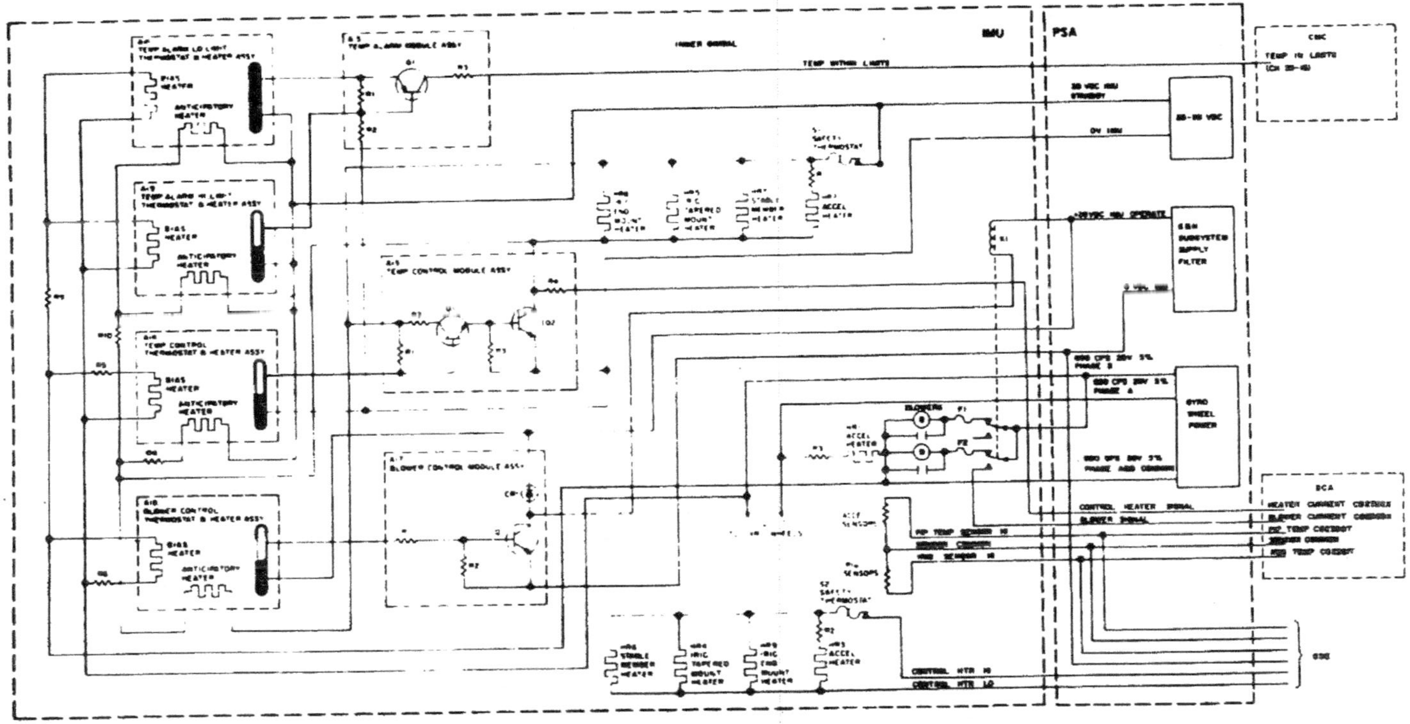

Fig. 4-28 Block II IMU Temperature Control System Schematic Design

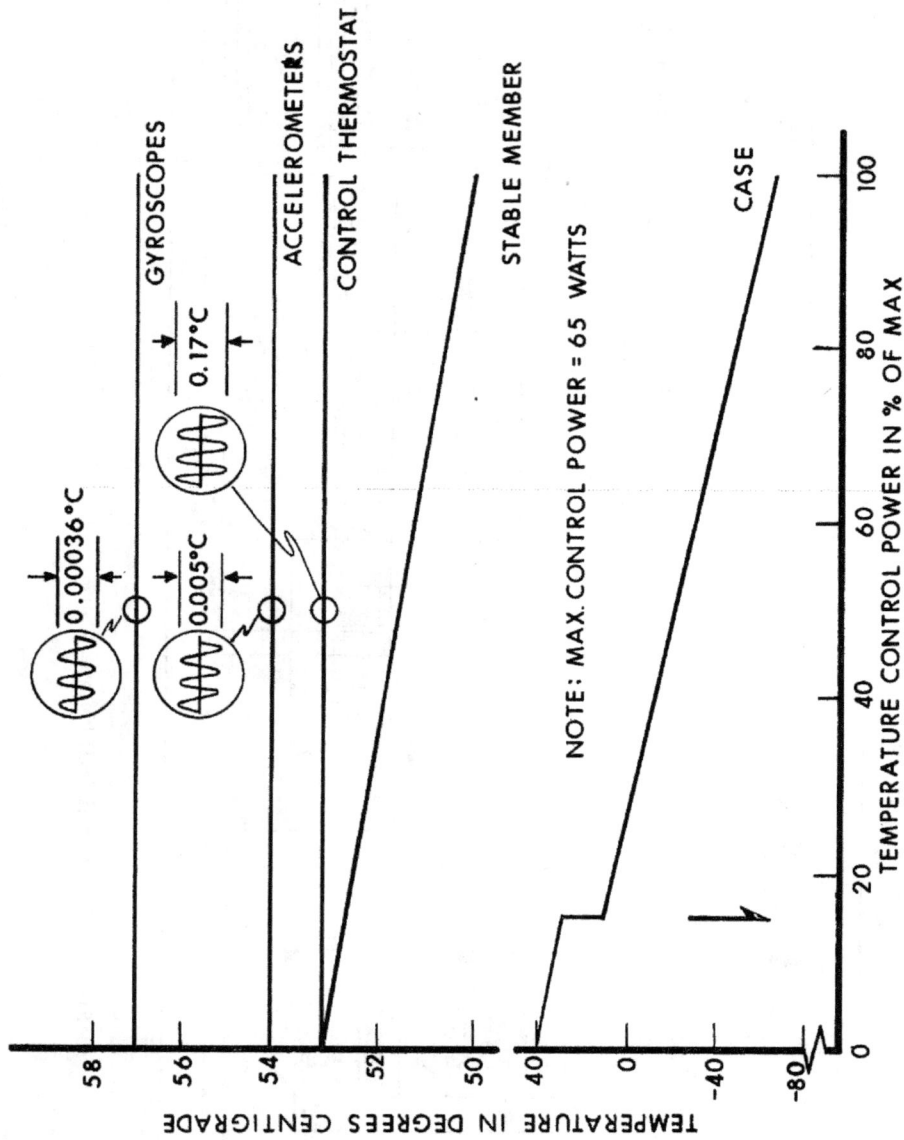

Fig. 4-29 IMU Temperature Profile vs Temperature Control Power

4.4.3  Mechanical Tests

4.4.3.1 Acceleration.  The Block I and Block II mechanical integrity IMUs were accelerated at 10, 20, and 30 g's along their outer axis.  These tests were run to determine the torque motor capabilities.

| Torque (% of max.) | IA | | MA | | OA | |
|---|---|---|---|---|---|---|
| | B1.I | B1.II | B1.I | B1.II | B1.I | B1.II |
| 1 g | 17 | - | 13 | - | 13 | - |
| 10 g's | 20 | - | 18 | - | 27 | - |
| 20 g's | 20 | - | 19 | - | 50 | - |
| 30 g's | 20 | 10 | 27 | 16.5 | 85 | 26 |

The gimbal breakaway torques were measured before and after these tests with no significant change measured.

4.4.3.2 Vibration & Shock.

(1) General -- Two mechanical evaluation IMUs were built for both Block I and Block II: a Vibration Model (IMU-VM and IMU 200-VM) and a Mechanical Integrity Model (IMU-2 and IMU-200-2).  The vibration models contained a dummy mass to simulate the stabilized member, and contained aluminum stub shafts and bearing mounts sized to simulate the stiffness of the final IMU design.  Tolerances were loose and no torque motors or resolvers were installed.  This model was used to determine resonant frequencies and transmissibilities.

The mechanical integrity IMUs were identical to functional IMUs with dummy inertial components that contained the vibration and acceleration pickups. These IMUs were used for vibration, acceleration, and shock tests to evaluate the adequacy of the mechanical design.

(2) Vibration Tests -- The IMU models were vibrated at 1, 2, & 3 g's (rms) sinusoidal input with a logarithmic frequency sweep from 20-2000-20 Hz in 16 minutes along each axis.  Each IMU was also vibrated with a 5 g (rms) random noise input, per the NAA specification, along each axis.  The results of these tests indicated resonant frequencies varying from 110 to 170 Hz with transmissibilities of 7 to 22.  A problem developed during the initial vibration tests of the Block I IMU and dampers were added.  This is discussed in Section 4.4.4.2.  The Block II IMU

vibration tests verified that the design was adequate. A Block II IMU was vibrated on the CM and LM navigation bases. The vibration input to the navigation base was in accordance with either the NAA or the GAEC random vibration specification. In each case, the input to the IMU was much less severe than the sinusoidal inputs used during the design evaluation tests.

(3) Shock Tests -- The Block II IMU was subjected to 20 g peak, sawtooth, 6 ms rise, and 1 ms decay time shocks twice along each axis with no problems developing.

4.4.3.3 Leak Tests. All IMU case parts were helium leak-checked prior to assembly. IMUs were pressurized to 24.7 psia for varying lengths of time and during the shock and vibration tests with no measurable change in pressure.

4.4.4 Design Problems

4.4.4.1 Changes in Materials. A brief review of the materials incorporated in the Block I and Block II Apollo IMUs showed that there were two that were new and different to this type of system. These were: (1) the friction material that was used in the vibration dampers of the Block I-100 systems, and (2) the Roll-Bond material used for both the Block I and Block II water-glycol case heat exchangers.

Vibration dampers were added to the Block I IMUs when a high vibration transmissibility was discovered during early tests. The selection of the type of damper and material was greatly influenced by the limited space available in the IMU. Some important characteristics that the damper material had to possess were (1) large dynamic coefficient of friction, (2) good wear resistance, (3) be very rigid, (4) stable under high temperatures, and (5) be machineable.

Johns Manville friction material style #230 was determined to be the optimum available material for this application. Final inspection of IMU-2, after vibration tests, revealed that the dampers and friction material had not degraded system performance.

An initial requirement for the Apollo IMU was that the fluid heat exchanger should be an integral part of the IMU. This led to the idea of using a roll-bond material (similar to that used in refrigerators) which had many advantages such as light weight, low thermal resistance, and required minimum volume in the CM and LM.

Techniques for roll-bonding of flat panels were well established. However, it was necessary to develop techniques to hydroform these panels into hemispheres, inflate the passages after hydroforming, and then machine and braze the parts to form the spherical IMU case. The material for the roll-bond parts was #6061 aluminum clad with #1100 aluminum to insure a good bond. During fabrication of the IMU case, a hydrostatic proof test was performed at 90 psig to insure that the parts were structurally sound.

4.4.4.2 Vibration Problems. As a result of high vibration magnification seen at the resonant frequency on the Block I IMU, fatigue cracks developed in the middle and outer axis stub shafts. Consequently, the stub shafts had to be redesigned and vibration dampers added to each axis in the torque motor inter-gimbal assemblies. Strain gage tests then indicated a reduction in the stub shaft stresses by a factor of 3. This, with the reduction of the transmissibility by the addition of dampers, resulted in a reduction of the stresses in the stub shafts to a level well below the fatigue limit of the material.

Various types of dampers were tested and a friction damper, which was the best for the space available, was added at the floated bearings.

4.4.4.3 Thermal Problems. The Block I temperature control system proved to be very reliable and presented very few problems after the design was completed. Considerable effort was expended in developing the magnetic amplifiers. Because weight was a major consideration in the overall G & N design, the early magnetic amplifiers were designed to be as compact as possible. This led to the use of very small wire sizes and resulted in failures from wire breakage. The amplifiers were redesigned in somewhat larger packages with heavier wire; consequently, no further failures occurred.

Another system problem was discovered during early system tests. The 3200 pps pulse-width modulated temperature control power was observed to generate 6400 pps spikes which could be seen throughout the system. An IMU wiring harness change reduced these spikes to tolerable levels.

The Block II system initially had several mercury thermostat failures. All of these failures indicated handling in excess of the specified 100 g's. A design change was made to incorporate a mercury thermostat with a 500 g, 2 millisecond duration specification, thus solving all breakage problems.

4.4.4.4 <u>Intergimbal Component Problems.</u> Four types of components were clustered about the bearings of the IMU: one gimbal-angle resolver transmitter, one torque motor, two slip-ring assemblies per axis, plus a single gyro-error resolver on the inner axis. The torque motors were completely trouble-free and need not be discussed further. The resolvers were subjected to minor changes in design. Only the slip-ring assemblies caused significant problems prior to qualification, but these had been well defined by the Block I experience; hence, Block II procurement was able to proceed with reasonable confidence. A more detailed study of the resolvers and slip-ring assemblies follows.

(1) Gimbal-Angle Resolver Transmitters -- The Block II gimbal-angle resolver was a dual pancake unit consisting of one-speed and sixteen-speed resolver transmitters wound on the same iron core. The two resolvers were wound as to have no mutual coupling; thus they functioned as separate resolvers while occupying little more space than a comparable sixteen-speed unit alone.

The principal problem in applying this resolver was attributed to the requirement imposed by the Block II CDU that the low side of the 16X sine and cosine winding be made common to ground. There was a distributed stray capacitance between the two secondary phases that caused a circulating current via the common low. A suitable equivalent circuit to illustrate the effects of the stray capacitance at 800 Hz is shown in Figure 4-30.

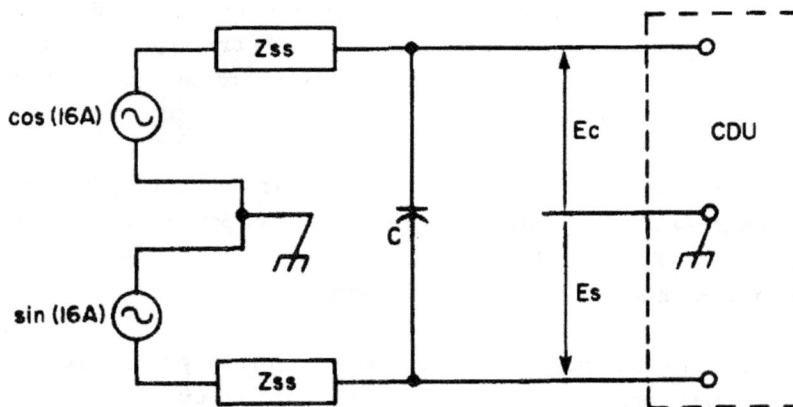

Fig. 4-30 Effects of Stray Capacitance

$Z_{SS}$ is the short -circuit output impedance of the resolver, and C is one-half the total capacitance between the secondary phases. Both an error in indicated angle and an additional quadrature voltage were impressed on the CDU input voltages by current flowing through capacitance C and the equivalent output impedances. Within a close approximation, both the error and the quadrature had the form cos (32A). Whereas errors of this form are usually corrected in high accuracy resolvers, there was no simple method to suppress this quadrature. As a result, a decision to wind the 16X secondaries for a step-down voltage ratio was made. The output impedance was greatly reduced and the stray capacitance reduced as well.

The other problematical feature of the Block II gimbal-angle resolver was increased errors as compared with the Block I design. This was essentially caused by the rather poor shape of the iron core required to minimize the IMU size. Another factor was less accurate (although within tolerance) grinding of the beryl-lium parts that mount the resolver as compared to the Block I parts (which had been subcontracted to a beryllium specialty house). In order to get reasonable yield, the vendor altered the winding pattern to further cancel some slot harmonics in the errors, and the specified maximum error was raised from 20 to 25 arc seconds.

(2) Gyro-Error Resolver -- The gyro error resolver was mounted on the inner axis, adjacent to the gimbal angle resolver. The error signals of the X and Z gyroscopes were rotated by this resolver into middle-gimbal coordinates, mak-ing the operation of the gimbal servos independent of the inner gimbal angle.

The principal constraints on the design of this resolver were:

a. It had to function with small signals when the gyros were near null and yet not saturate before the gyro preamplifiers did. This requirement necessitated an iron core with excellent linearity.

b. It had to be shielded from the adjacent gimbal-angle resolver.

c. Its axial width including the shields had to be small to conserve space and weight.

d. Its impedance level had to be a reasonable compromise between loading effects on the gyro preamplifiers, and its contribution to source impedance for the wiring capacitance and demodulator loading. In this regard, one must note that the demodulator load was a function of applied voltage. Also observe that both signal phase shift as well as magnitude had to be maintained across the resolver.

e. An 0.003 inch minimum air gap was specified for the resolver to insure against mechanical interference resulting from tolerance accumulation and stress effects.

f. The shields influenced the performance of both the gyro error and the gimbal angle resolver.

The difficult design problem of the inner-axis resolvers was attenuating the 3200 Hz cross-coupling between them. Only one vendor achieved the specifications and qualified as a source. Because the shield design would be much easier with more axial space, this problem interacted strongly with other parameters which could have been improved with more space. In the meantime, those responsible for stabilization loop design were evaluating the gyro error resolver design based on a "figure of merit" consisting of the ratio of tuned open circuit input impedance to the magnitude of the short-circuit output impedance. Such a figure, being based on effects in the stabilization loop circuits only, did not weigh the more complex aspects of space and cross-coupling.

Although prototype gyro error resolvers successfully met all the specifications, a difficulty developed in production. Impedances and the "figure of merit" were highly sensitive to both air gap length and lamination magnetic properties. In early 1966, a position was reached where almost no yield was obtained that met the minimum air gap and also the required impedance. To keep on schedule, a few units were accepted to a revision of the specification calling for a minimum air gap of 2.2 rather than 3.0 mils. This was made possible without reducing the minimum running clearance by performing the final grinds with the laminations already mounted in the stub shaft and bearing mount. Meanwhile, a new winding was designed which, with a 3-mil air gap, achieved rather different but quite acceptable impedances.

(3) Slip-Ring Assemblies -- The Block II slip-ring assemblies never actually caused significant problems. This was true, however, only because a number of steps taken to achieve qualification of the Block I assemblies paved the way for Block II. The ensuing discussion will therefore be a short review of Block I problems, solutions, and resulting effects on Block II.

The early Apollo experience with slip rings was based on assemblies from Electro-Tec, Poly Scientific, and Collectron Corporations. Between experience in prototype IMUs and attempts to qualify all three sources, an impressive list of failures was accumulated. Contamination, bearing failures, high noise, i.e., resistance variation with rotation, and various minor mechanical problems were experienced. Poly Scientific Corporation was eliminated when they were unable to cure a contamination problem after many attempts. Collectron had trouble maintaining bearing preload but was eliminated mainly because of high wear and noise early in all life tests. This left Electro Tec, who had experienced numerous small bearing failures caused by contamination after exposure to a severe humidity cycle. Electro Tec also had some out-of-tolerance noise at the conclusion of the life test.

The prospects of the Electro Tec units actually were found to be good. They had experienced difficulty only with the smaller bearing. Their Block II design used two of the larger size, however. Furthermore, on teardown, most bearing failures were found to be the result of improper lubrication. Consequently, Electro Tec began buying the bearing pre-lubricated with good results. Next, a humidity cycle was specified which more realistically represented the actual conditions of storage and use, and the contamination failures ceased. Finally it was recognized that the life test, or, officially, Extended Performance Test, was conducted with oscillations and slews which gave rotational rates up to three times those in an IMU under a locked-gimbal tumble. After changing to a more realistic activity level during the test, Electro Tec slip ring assemblies completed life testing within the specified noise limits. (The 10 cpm + $360^\circ$/ $-360^\circ$ oscillation was changed to 2 cps, $3^\circ \pm 1^\circ$ double amplitude dither, superimposed over a 1/2 to 2 rpm rotation.)

These noise limits, more nearly, although still severe, simulated the conditions expected in testing and operation of the IMU. The actual slip ring motions in an IMU were impossible to duplicate exactly since they depended on test or flight conditions. For a large percent of the time during tests or flight the brushes were not moving over the rings since the IMU was stabilized or was moving at meru rates as a result of gyro drift. Periodically, the brushes were rotated to a new position when the IMU was rotated to a new position by the test fixture or the spacecraft. As the axes arrived at a new orientation, there were some minute (seconds of arc) random oscillations as the IMU nulled itself at the new location. The normal wear of the slip ring in the IMU was caused by random oscillations. The slow rpm was superimposed so that all areas of the ring were checked since the location of the brushes on the ring varied from slip ring to slip ring in the IMU. A relatively high rpm test was included every 100 hours to simulate the condition during tests where a malfunction in the system would cause a gimbal axis to rotate with full torque from the torque motors. The 600 rpm requirement in the original test was in error and was changed to 200 rpm since the IMU was limited to less than this rate because of friction and motor back emf.

The results were that Electro Tec Corporation was the only qualified supplier of Block I slip ring assemblies and became the principal supplier for Block II while other manufacturers attempted to cure their difficulties and qualify for Block II.

## 4.5    Coupling Display Unit (CDU)

### 4.5.1    Block I CDU

The Block I coupling display units provided a coupling between the Apollo Guidance Computer (AGC) or the astronaut and the IMU and the optics. Either the AGC or the astronaut could have transmitted angular data to the IMU. The CDU also provided the astronaut with a visual readout of the IMU and optics angular data during the coarse align, fine align, and manual CDU modes of operation. Desired angles were transmitted by the CDU loop to the IMU in analog form during the coarse align and manual CDU modes. The CDU loop was also capable of repeating the IMU gimbal angles and the optics angles, and transmitting this angular data to the AGC in digital form in the fine align mode. The CDUs in the ISS could also have been used to provide a coupling between the astronaut or AGC and the spacecraft autopilot. Five CDUs were required by the G&N system, three in the ISS, and two in the optical subsystem (OSS). Each CDU contained a servo motor-tachometer, four resolvers, a digital encoder, three display dials, and a thumbwheel; all were interconnected by a stainless steel gear train. Each CDU also contained a slew switch, which was not utilized in the OSS CDU loop application. The gear train was driven by either the thumbwheel or the servo motor-tachometer. The servo motor-tachometer was an 800 cps unit that drove the CDU gear train and produced a feedback signal proportional to the output shaft rate. The motor was driven by the Motor Drive Amplifier (MDA) in the PSA. Input signals to the MDA were received from the CDU resolvers, the digital-to-analog converter in the PSA, and the slew switch, all of which represented CDU loop errors. The motor converted the various CDU loop error signals into the appropriate output shaft angle and velocity. The tachometer provided a negative feedback signal proportional to the motor shaft speed to the MDA. The slew switch provided a 6.25 Volt, 800 Hz signal of zero or pi phase to the MDA to position the ISS CDUs during the manual CDU mode of operation. The four resolvers were utilized either for angular data transmission or angular error resolution. The specific resolvers utilized and their functions depend on the CDU application. The digital encoder (Figure 4-31) converted the angular motion of the CDU output shaft into a digital signal which could be used by the AGC.

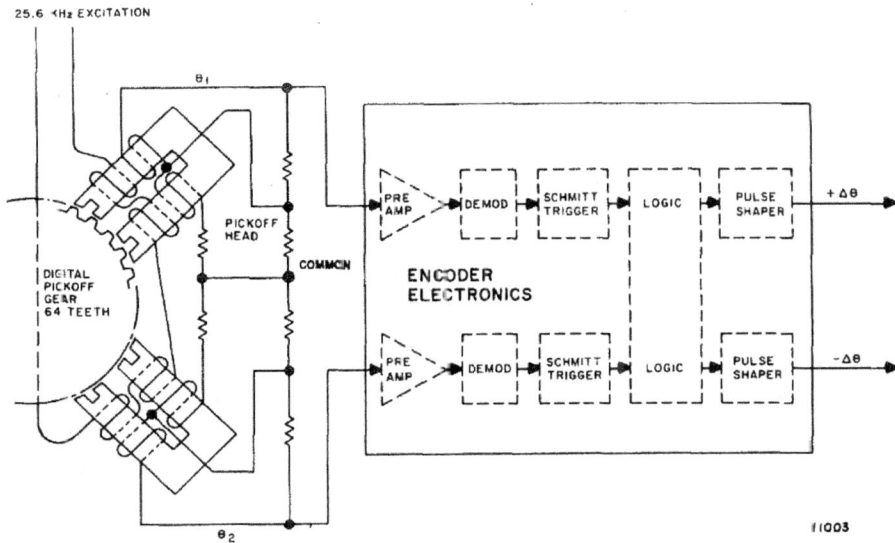

Fig. 4-31 Digital Encoder Functional Diagram

The digital encoder consisted of a digital pickoff gear, two U-shaped digital pickoff heads, and the encoder electronics. Primary and secondary coils were wound on each of the two legs of each pickoff head. The primary coil was excited with 25.6 k Hz. The secondary coils were connected in opposition. When the movable armature (pickoff gear) was symmetrically located between each of the two legs, the mutual inductance between the two legs was the same. Since the secondaries were wound in opposition, no output voltage resulted. This position of the gear was referred to as the null position. If the armature was displaced from the null position, a greater voltage was induced in one of the coils. Motion of the gear tooth in the opposite direction similarly produced a greater output voltage in the other coil. The output voltage varied sinusoidally with the displacement of the gear teeth. Thus, this device modulated the excitation at a frequency proportional to the velocity of the armature. The second pickoff head was located so that its output voltage would be displaced from the first output voltage by 90 degrees. The relative phasing between the two outputs indicated the direction of shaft movement.

The first portion of the digital encoder electronics was a preamplifier which was followed by a synchronous demodulator. The output of the demodulator was converted into square waves by the Schmitt trigger. The square wave outputs were then fed into a logic network which identified the direction of rotation and produced voltage spikes on one output line ($+ \Delta \phi$) for clockwise rotation of shaft and on the other line ($-\Delta \phi$) for counterclockwise rotation. These spikes were then converted to the proper logic level for the AGC by pulse shaping networks. Each pulse presented to the AGC represented 40 arc-seconds of shaft travel when used with the ISS, or 20 arc-seconds when used with the optics. The logic of the encoder electronics provided two distinct encoder modes of operation, either one pulse per gear tooth or two pulses per gear tooth. Again, the modes of operation depended on subsystem mechanization.

While the digital encoder was providing the AGC with information about the position of the CDU, the AGC had also to be provided with a means of controlling the movement of the CDU. This was accomplished by the digital-to-analog converter (DAC). The DAC received commands from the AGC and converted these commands to an electrical output usable in the MDA. The inputs from the AGC were pulses of approximately 3 microseconds duration with a repetition rate of 3,200 pps. These pulses were not applied continuously but were supplied in bursts from the AGC. The feedback signal from the encoder also consisted of pulses, but the repetition rate varied according to the speed of encoder pickoff gear rotation. The AGC inputs caused a capacitor in the DAC to be charged. The charge on this capacitor was reflected into the DAC output stage by two integrated choppers operating at 800 Hz The output of the DAC was an 800 Hz square wave with a maximum amplitude of approximately 10 Volts peak-to-peak. This output was utilized by the motor drive amplifier to position the CDU. The negative feedback from the encoder was required to insure that only a defined amount of CDU movement per pulse would occur from the AGC. The encoder input discharged the capacitor that had previously been charged by the AGC input, therebyshutting off the DAC.

4.5.2  Block II

The Block II coupling data unit comprised the central angle junction box between the IMU, optics, computer and the certain portions of the spacecraft analog electrical interfaces. There were three basic portions of the CDU: the angle read system or analog-to-digital conversion process, the digital-to-analog conversion process, and a portion of the moding controls for the guidance system. The analog-to-digital system is covered in some detail while the other portions are only briefly described.

4.5.2.1 Analog-to-digital Conversion. Angle information was stored in a two-speed resolver system of a control member, for example, a gimbal axis or an optical axis (see Figure 4-32). The output of the resolver was proportional to $\sin \theta$, $\cos \theta$, $\cos \theta$, $\sin \eta \theta$, $\cos \eta \theta$, where $\eta$ was a binary number. It was a common technique to have both the single speed and multiple speed resolver use the same iron and utilize a single excitation winding. The second excitation winding space, phased $90^\circ$ with respect to the primary excitation, was used for electrical zero adjustment. The elements of the angle read system were an analog multiplication of the resolver output, an analog summation, a sampler and quantizer, a storage counter to control the analog multiplication, and ac switches controlled by the counter to gate inputs to the analog multiplier.

The equation mechanized was:

$$\sin \theta \, \cos \psi - \cos \theta \, \sin \psi = \sin (\theta - \psi)$$
$$\cos \theta \, \cos \psi + \sin \theta \, \sin \psi = \cos (\theta - \psi)$$

where $\psi$ is a quantized angle in increments of 22-1/2 electrical degrees, and $\theta$ is the angle of the control member.

As shown in Figure 4-33, this output was summed with a quantized linear interpolation of difference between $\sin (\theta - \psi)$ and $\theta$. The selection of the quantized angle and the quantized linear interpolation angle $\phi$, was based upon the contents of an angle counter register. The angle counter inputs were gated by the phase of the summed voltages. Thus when the contents of the counter were equal to the control member angle:

$$\sin (\theta - \psi) + K \cos (\theta - \psi) = 0$$

The inputs to the counter were angle increments of the control member and these were parallel-fed to the computer where the same control member angle information was stored.

Analog multiplication of the resolver $\sin \theta$ and $\cos \theta$ voltage was accomplished by the use of an ac operational amplifier with the ratio of the feedback resistor to the input register equal to the cosine of the angle $\psi_1$. S was an ac transistorized switch gated closed or opened by contents of the angle counter resistor. The switch was in series with the feedback resistor to take advantage of the high impedance open condition and low impedance closed condition which made the impedance

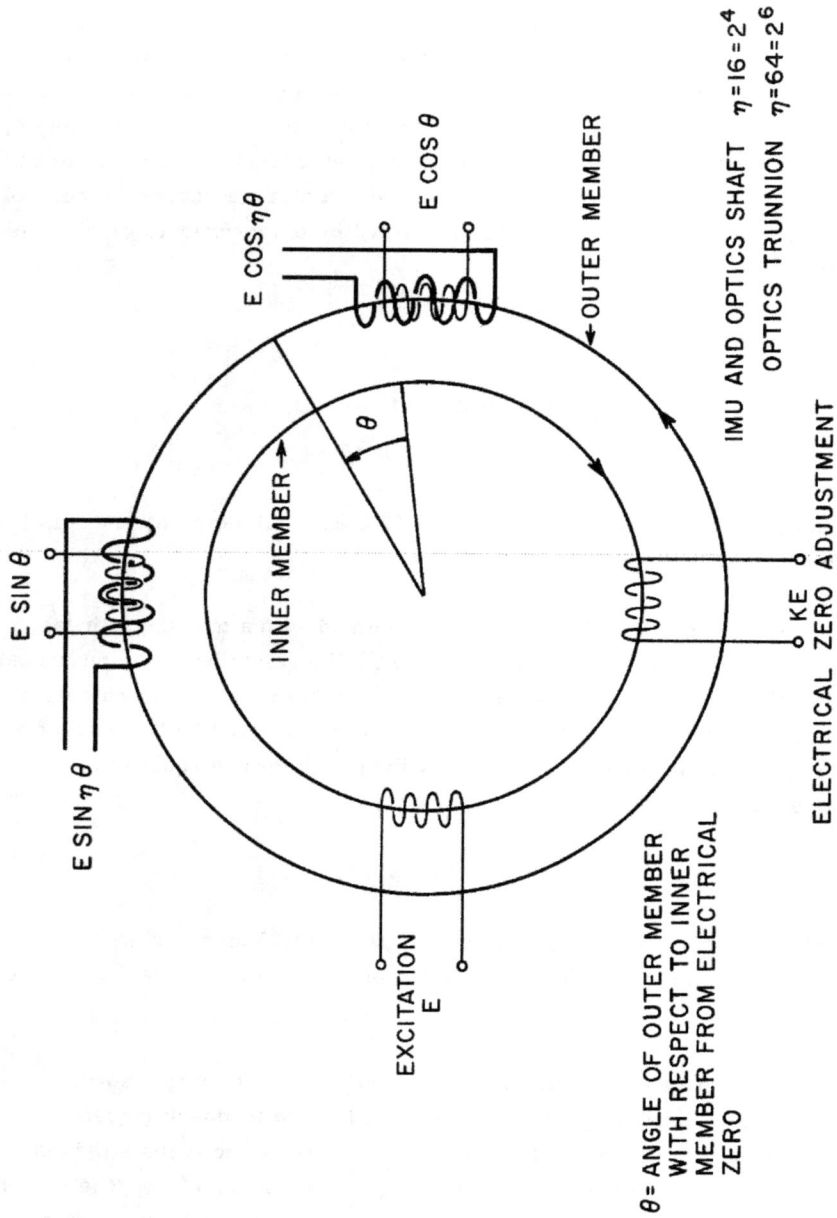

Fig. 4-32 Resolver Schematic

E COS $\eta\theta$

E COS $\theta$

OUTER MEMBER

IMU AND OPTICS SHAFT $\eta=16=2^4$

OPTICS TRUNNION $\eta=64=2^6$

E SIN $\theta$

INNER MEMBER

$\theta$

E SIN $\eta\theta$

KE

ELECTRICAL ZERO ADJUSTMENT

EXCITATION E

$\theta$= ANGLE OF OUTER MEMBER WITH RESPECT TO INNER MEMBER FROM ELECTRICAL ZERO

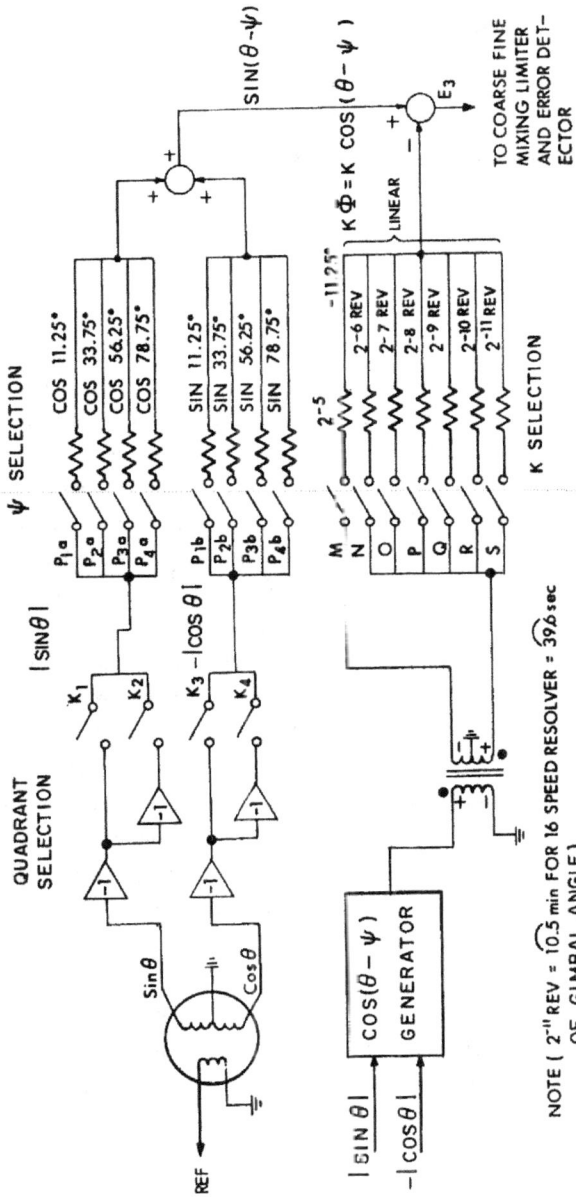

Fig. 4-33   Selection Logic - 16 Speed Resolver Digitizing Loop

$A \cos \psi_1$

$\sin \theta \cos \psi_1$

$\sin \theta$ —— $A$ —— $K$ —— $S$

$A \cos \psi_2$

$0$

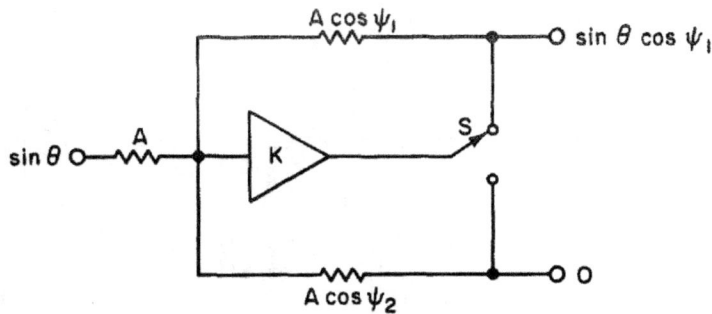

equivalent to a portion of the amplifier gain, K. Since the switch was effectively a single-pole double-throw switch, the output of the open side differed from zero only by the input signal divided by the amplifier gain. Using the technique of transistorized ac switches and operational amplifiers, the read system was mechanized. The read counter contained 16 bits. The lowest order bit was used to eliminate transmitting any limit cycle operation to the computer and thus avoided creating unnecessary activity. The four highest order bits were used for quadrant selection and multiplication of the single speed resolver ($\psi_1$). In addition, the bits $2^9$ - $2^{11}$ were used as an approximate linear interpolation of the single speed resolver to within $2.81°$ of the actual angle. The multiple speed resolver (sixteen speed) was the precision angle transmitting device. The zero-to-peak errors of this resolver were less than 10 seconds of arc. There was crossover between the sixteen speed and the single speed resolvers to assure synchronization of the reading of sixteen speed resolver within the proper cycle. The lowest order bits were a linear interpolation of error using the voltage of the cos ($\theta - \psi$) as a source. This voltage had the same phase relation as the sin ($\theta - \psi$) of the sixteen speed resolver and was scaled correctly by the resolver attenuation.

Referring to Figure 4-34, the input to the error detector was the sum of the single speed multiplication, the sixteen speed multiplication and the linear interpolation. There was a coarse-fine mixing network to assure synchronization and angle measurement using the precision resolver. The error detector contained an active feedback quadrature rejector network which for large error signals would not introduce dynamic errors for reading the angle, but for small errors would yield the proper precision. The output of the error detector was fed to both the rate selection logic and up-down counter logic. The contents of the counter were used to control the ac switches for the multiplication of the resolver voltages and the linear interpolator.

The error detector had three-state or ternary logic. The lowest order pulse rate command to the counter was 800 pulses per second. Using this as the lowest order assured switching of equal multiples of the resolver carrier frequency, 800 Hz, which prevented rectification of the switched signal and altering of the dynamic operation of the read system.

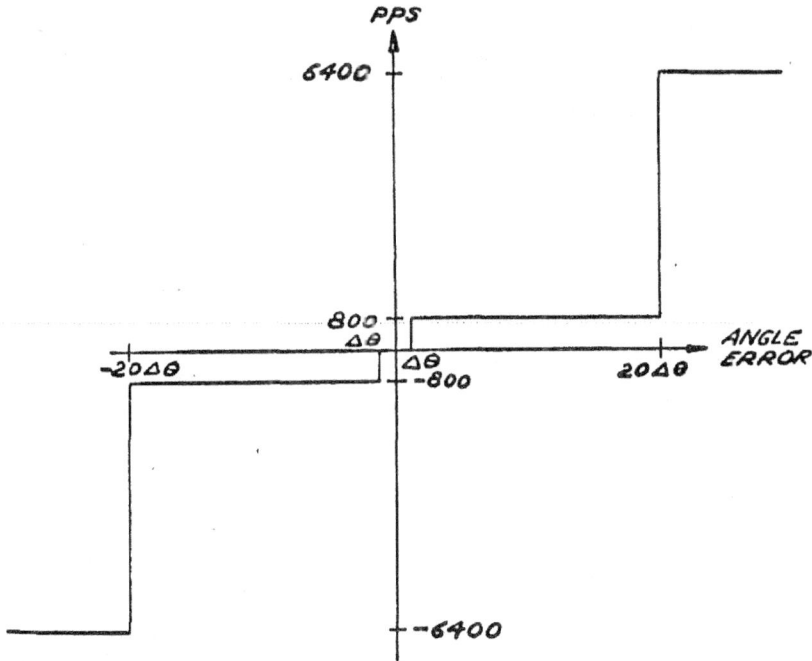

The high speed rate following command reduced dynamic error for high angular velocity inputs and the low speed command rate reduced the limit cycle error.

The linear interpolation constant K was adjusted to minimize the peak error

$$E = \sin(\theta - \psi) - K \cos(\theta - \psi)$$

By suitably choosing K, the error for the speed resolver system could be reduced to less than 10 arcseconds. For a 64 speed system as used on the optics trunnion, this error was reduced to less than 2 arcseconds.

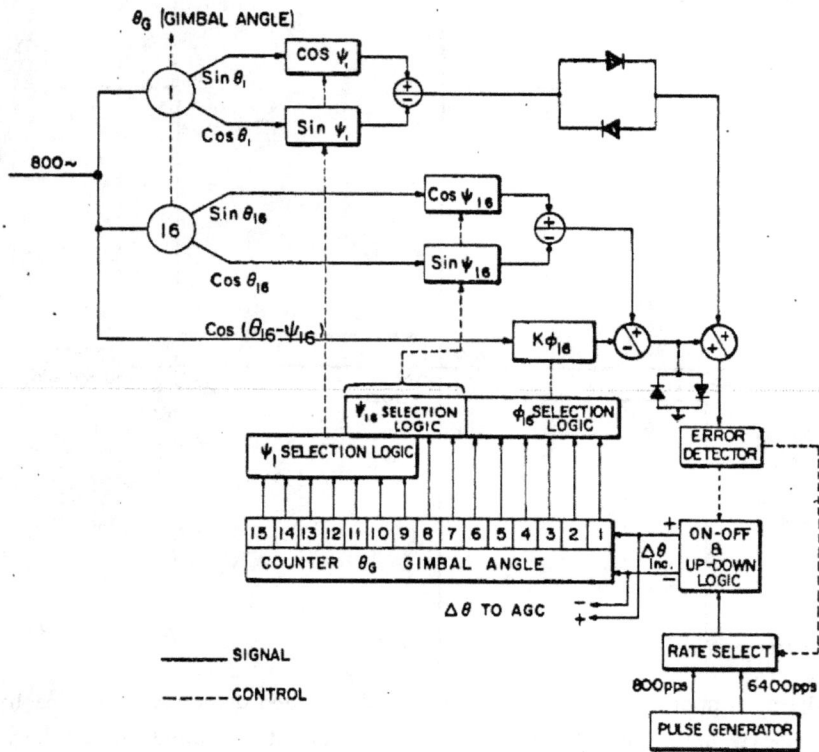

Fig. 4-34 Coarse-fine Mixing (1 and 16 Speed Resolver)

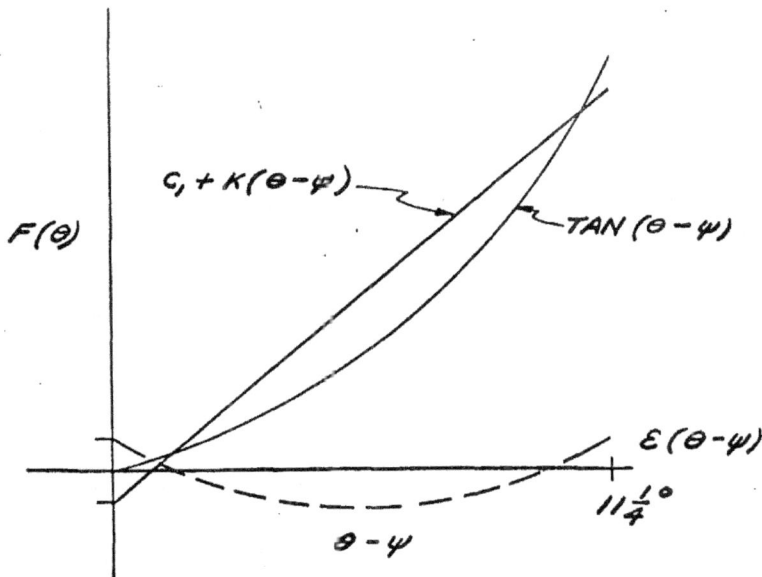

In addition, a bias was added to further reduce the errors over the range of linear interpolation. This resulted in a system with errors within the range of predicted errors.

All digital functions including the memory were mechanized with a three input NOR gate, a silicon semiconductor micro-integrated circuit. This was the same element used in the computer. Direct-coupled transistorized logic was used throughout. A multiple phase clock system, generated within the CDU and synchronized to the guidance computer, was used to control all functions.

4.5.2.2 Moding and Digital-to-analog Conversion. The guidance computer served as the important link between the spacecraft and the sensing device. All angle transformations were made by the guidance computer based upon IMU gimbal angles or optics angles. For any steering function utilizing the IMU, the computer through the use of knowledge of the gimbal angles provided steering and attitude commands to the spacecraft via the digital-to-analog converter. The angle information was always stored in the resolvers, and no mechanical rezeroing of the IMU was necessary to establish within the counter the IMU gimbal angle within the uncertainty of the resolver, the error of the read system and the bit size of the analog-to-digital converter.

The digital-to-analog converter system was required to accept digital commands from the guidance computer and generate analog voltages (ac and dc) proportional to these commands. The error angle counter was an eight bit up-down counter. Computer commands were stored in this counter. The counter was logically controlled to prevent it from being reset to zero in the event it was commanded to more than $10^9$ increments. The analog error signals were developed by an 800-Hz source gated by the counter contents through a resistance ladder to the input of an operational amplifier. For polarity, reversal by switching was used to command the 0 phase or $\pi$ phase 800-Hz input to the ladder. All analog voltages used as steering commands to the LEM, Spacecraft or other portions of the Apollo launch vehicle were dc-isolated from the guidance system, either by transformers or an isolated demodulator.

The moding of the system was almost entirely controlled by the computer. Provision was made for two modes to be controlled by the astronaut. One mode was the cage function. In the event of a spacecraft tumble and loss of attitude, provision was made for the astronaut to cage the IMU gimbals with respect to the spacecraft after he has first stabilized the craft. This then gave him a means of obtaining a new reference in a very short time. The other manual mode was similar in that with the computer not operating he could use the same switch to cage the IMU with respect to the spacecraft, release it, and again use the IMU as an attitude reference.

In all other inertial subsystems, moding was controlled by the computer. There were a number of interesting modes possible because of the flexibility of the CDU. The Coarse Align Mode, that of commanding the IMU gimbal angles by the resolvers, was rate controlled to limit angular velocity input to the gyroscopes. The input to the error detector was summed with the analog command from the computer to provide stable operation. The gimbal angle pulse increments from the read system were used as feedback pulses to the error angle counter.

### 4.5.3 CDU Problems
4.5.3.1 Block I Problems. The problems associated with the Block I CDU were of a mechanical nature. The gear trains used with the CDU exhibited excessive wear, and a few units "froze" in operation. To correct this failure mode, a carefully selected lubricant was added to the gears. Another gear box associated failure was with the motor tachometer supplied by one of the two vendors of this component. Because of mechanical tolerances, it "froze" at elevated temperatures. The corrective action for this failure was to select the motor-tachometer from the vendor whose product did not exhibit this failure mode.

4.5.3.2 Block II Problems  The operation of the CDU, in a system configuration, disclosed two problems that required design changes.

(1) The coarse-fine crossover for the CDU was originally designed to take place at a maximum coarse error of 3 degrees. System tests showed that under certain conditions an oscillatory limit cycle condition developed between the coarse and fine systems. A design change was made to reduce the crossover point to 7.5 degrees from null. This solved the problem.

(2) The CDU contained capacitor-coupled transistor switches as shown. Immediately following power application or after long periods of inactivity, the dc charge that would normally accumulate on the capacitor would not be present or would have "leaked off."

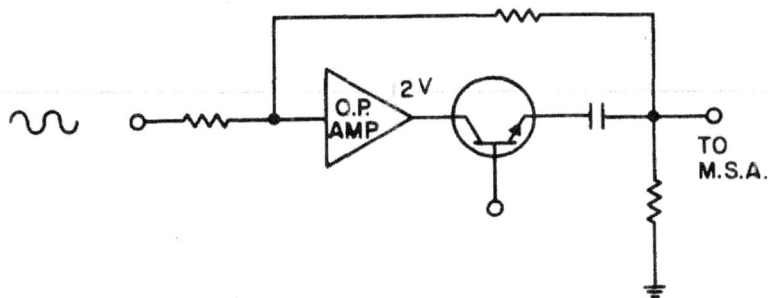

When switching action was initiated, the charging action of the coupling capacitor in conjunction with the frequency response of the operational amplifier would create a low frequency "bounce" on the output of the switch. The cumulative effect of several switches being activated during coarse alignment of the IMU would cause the input limiting diodes of the error amplifier to be driven into the active region and effectively short out the ac error signal to the amplifier. See following illustration.

ERROR
AMP

LIMITING
DIODES

TO
TERNARIES

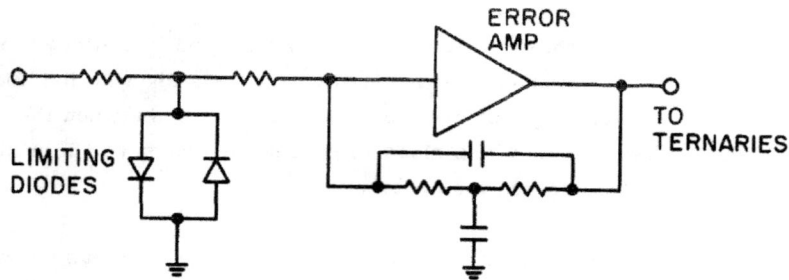

With the loss of this error signal, the read counter would not increment and the feedback pulses to the DAC error counter would not be present. The AGC would "load" the DAC error counter to perform the coarse alignment, and in the absence of the feedback pulses to subtract the "loaded" angle, when the gimbals moved, the error counter would overflow on subsequent computer commands producing a loss of information to the CDU and a failure to achieve the commanded CDU angle.

The diodes were removed and the error amplifier modified to improve its saturation characteristics thus producing a larger linear operational region.

(3) The Block II CDU encountered several component associated problems during the manufacturing period of the Apollo program. The part types that exhibited failure modes were (1) RTL micrologic, (2) transistors, (3) transformers, (4) capacitor, (5) relays, and (6) resistors.

a. The micrologic NOR gates used in the digital modules of the CDU exhibited both "open" gate failures and "shorted" gate failures during subassembly testing at AC Electronics Corporation. The majority of the failures occurred during vibration tests. A minority of the failures were attributed to over-voltage stress applied by the test fixture during the early stages of the program.

The failures induced by the vibration of the modules were attributed to contaminants within the flatpack of the micrologic. A "screening" vibration test was conducted at the module manufacturer to remove this mode of failure. The vibration "screening" vibrated the micrologic flatpacks along three perpendicular axes.

b. The transistor used in the CDU ac switches was a 2N2351 device. During the program this transistor exhibited failure modes that were attributed to the "purple plague" and to internal contamination. The corrective action taken was to institute a centrifuge and X-ray inspection for incoming transistors at AC Electronics and to improve cleaning procedures at the manufacturer's facility. A second source of supply was also obtained for parts procurement.

c. Transformer failures occurred via two different modes. The Bush transformer (1010724) used in the main summing amplifier exhibited inductance shifts after potting the module. The inductance shift was found to be the result of lamination shifts within the transformer. This condition was corrected by an improved method of sealing the laminations.

The other mode of transformer failure was associated with wire breakage internal to the UTC transformer. The wire size used to wind these transformers was #50 AWG. The internal stresses placed upon the wire terminations and the fine wire used to wind the transformer were believed to be caused by the hard potting compound used in the transformer. The corrective action was to replace the transformer with a compatible unit wound with larger wire.

d. Failure modes of the capacitors used in the CDU were associated with two types of capacitors. One capacitor exhibited high leakage characteristics after potting in a module and after being subjected to vibration testing. This failure mode was found to be predominantly associated with a capacitor from one manufacturer (Kemet). The corrective action was not to use this product in the construction of the CDU. The other capacitor was a polystyrene film unit. This capacitor exhibited film rupture resulting in shorted units and poor connections to the film for external leads. The corrective action was to replace the polystyrene unit with a Mylar capacitor.

e. The relays used in the CDU for DAC output transfers were type SCD1010353-7. Contamination of the relay by solder balls produced failures of these units in the initial stages of CDU manufacturing. Subsequent tightening of the inspection procedure by the manufacturer removed this failure mode as a problem.

f. Failures associated with resistors were confined to the high resistance value metal films and the carbon resistors used in the main summing amplifiers. The metal film resistors were found to be changing values under module rework. It was concluded that the abrasive material used during depotting was establishing a high electrostatic voltage on the resistors and in turn punching through and/or changing its metal film characteristics. The corrective action was to change the type of abrasive material used during the depotting process.

The carbon resistors, of which there were only two in each CDU axis, exhibited a drift characteristic upon potting that forced the associated resistance value to fall outside the specified limits. The corrective action for this phenomenum was to change the specification to accommodate the drift. These resistors occupied a noncritical position in the function of the CDU; hence, the specification variance could be tolerated.

## 4.6    Electronics Packaging

### 4.6.1    Constraints on Packaging Design

The G&N electronic packaging design evolved as the result of compromises among conflicting constraints. Besides the principal goal of achieving required performance within the operational environment, the impact of schedule, weight, reliability, cost, power drain, producibility, and logistics were all considered. The constraint of schedule affected both design and design evaluation. Schedule requirements dictated the early release of the design to production. Functional, mechanical and thermal design evaluation of prototype hardware in the expected environment occurred simultaneously with early production. Necessary design modifications resulting from the evaluation were sometimes incorporated as repair fixes.

Factors which had primary influence in the packaging design were the spacecraft environment and interface constraints, and for Block I, the requirement for inflight diagnosis and module replacement.

### 4.6.2    Interface Control Documents

Spacecraft constraints were defined in Interface Control Documents (ICD). The ICDs affecting the package design defined volume envelopes, weight requirements, thermal interfaces, mounting arrangements, acceleration, vibration and shock environments, material compatibility, and color and markings. The ICDs developed as the G&N and spacecraft designs matured. Early gross interface definitions were refined with occasional design modifications on both sides of the interfaces to accommodate other associated designs.

4.6.3 Inflight Maintenance

The requirement for inflight maintenance was a consequence of the high reliability requirements imposed by NASA on all the Apollo contractors. Since the failure rates of the high reliability components for Apollo had yet to be determined, the conservative approach was to provide for inflight maintenance.

Block I experience, however, showed that the required overall reliability could indeed be realized without resorting to inflight maintenance. This requirement was therefore removed for the Block II command module and LM G&N systems.

4.6.4 Materials Testing

NASA and all contractors had been doing extensive flammability, toxicity, outgassing, and odor testing of materials in accordance with NASA "Procedures and Requirements for the Evaluation of Space Nonmetallic Materials" MSC-A-D-66-3. As a result of these tests, many engineering materials were banned from use within the spacecraft if they were to interface with the cabin atmosphere. Also, in Block II, a great emphasis was placed on avoiding the use of flammable materials. These considerations led to the decision to package the equipment, whenever possible, into sealed metal containers. Because the Block II and LM inertial subsystems were functionally identical, an attempt was made to use common equipment wherever possible in the two systems.

4.6.4.1 Corrosion Protection & Associated Design Restraints. The extensive use of magnesium in the structural members of the Apollo airborne electronic assemblies introduced the possibility of serious corrosion problems that required solution. There were certain parameters common to all the assemblies that tended to compound the problem when considering methods of protecting the equipment against corrosion. In addition to basic protection, the requirements for both the Block II CM and LM finish systems were: (1) electrical grounding, (2) thermal transfer, and (3) outgassing. A fourth consideration, for LM assemblies only, was emissivity.

Hard vacuum $\left[\text{vacuum below the level of } 10^{-6} \text{ mm Hg (torr)}\right]$ affects materials in two general ways: First, the material contains adsorbed and absorbed gasses which may be removed by outgassing. Secondly, vaporization or sublimation of the material or of a volatile (or toxic) component of the material may occur. The rates of these reactions can be radically affected by temperature. The amount and nature of potentially toxic and objectionable products outgassing from crew bay materials were evaluated and analyzed to insure astronaut safety and well-being for optimum mission performance. The outgassing concentration of individual contaminants was related to the effective habitable spacecraft volume. In general, materials that outgassed organic substances in excess of 100 ppm per weight of sample was considered unacceptable for use in manned spacecraft crew bay areas.

The selection of organic materials for corrosion protection was evaluated and considered acceptable in terms of outgassing characteristics.

Outgassing products from the modules were entrapped in the hermetically sealed containers. Outgassing contaminants from the wire-wrap side of the headers were minimized since this was the side that mounted directly to a coldplate and/or spacecraft structure.

4.6.4.2 Required Surface Treatments. Electrical bonding of the PSA, PEA, PTA, and CDU to spacecraft structure was accomplished by using a spotface under mounting screw heads as an electrical conductive path. The engaged threaded section of the mounting screw completed the path to the spacecraft structure ground. Thus electrical currents were flowing through the case to the S/C structure via a dissimilar metal path. The total number of electronic assembly mounting screw fasteners was analogous to a parallel grounding circuit for a given assembly. Each assembly-to-structure electrical resistance was to be less than ten milliohms. Dissimilar metals were an inherent part of the design, and the only means of controlling galvanic corrosion was to exclude moisture from the mated metal surfaces. The solutions for the various assemblies were as follows:

(1) CM PIPA Electronics Assembly (PEA) -- A stepped, rubber gasketed iridite aluminum washer was used under the cover screws, or as an alternate, the DOW 19 spotface was coated with an RTV-108 after assembly of the cover-to-header screws.

(2) CM Power and Servo Assembly (PSA) -- The counterbores in the cover were filled with an RTV-108 after assembly.

(3) CM and LM Coupling Data Unit (CDU) -- (a) The cover screws and nuts, after assembly of the two halves, were coated with RTV-108; (b) the lower washer, which was under the captivated header-to-structure mounting screw head, was bonded in place. The adhesive surrounded the washer periphery, masking the Dow-19 magnesium surface. Epon 919 was used as an adhesive.

(4) LM Power Servo Assembly (PSA) -- (a) A stepped, rubber gasketed iridite aluminum washer was used under the cover screws; (b) a thin bushing sleeve was bonded to the magnesium in the precessed wells.

(5) LM Pulse Torquing Assembly (PTA) -- Same as item (4).

The electronic assemblies required a means of dissipating generated heat. The modules transferred heat to the assembly header which in turn transmitted heat to a coldplate heat exchanger or the spacecraft structure. The assembly-to-coldplate or structure interface surfaces were Dow-17 magnesium with zinc chromate primer (MIL-P-8585) and 3M 400 polyurethane paint (SCD 1012543) (per ND1002291).

Total primer and paint thickness was 2 mils maximum. A coating was required on these magnesium surfaces, for the mating surface was alodine aluminum. This meant that thermal hot and cold spots would have condensed water, creating a corrosive galvanic couple. This coating reduced the thermal transfer efficiency when compared to an unpainted interface, but it was within the limits to maintain sufficient cooling even in a vacuum environment. This was partially accounted for by the application of DC-340 thermal grease to the assembly mounting interface for the CM and the application of alumina-filled RTV-601 for the LM.

External surfaces, with the exception of heatsink surfaces of actively cooled electronic assemblies in the LM, possessed an infrared emittance of 0.40 or less in the region of the electromagnetic spectrum between 5 microns and 35 microns. This was a requirement for thermodynamic equilibrium in the LM. A number of finishes were evaluated using an emissometer, and it was determined that an aluminum filled epoxy paint developed by MIT/IL provided an emittance value around 0.30 when applied on Dow-17 magnesium.

The specific requirements for Apollo electronic equipment with regard to low emissivity, outgassing, thermal conductivity, EMI shielding and corrosive environments led to studies in these specific areas. It is beyond the scope of this report to describe all the surface treatments, conversion coatings, organic coatings, or any combination of these treatments that were tested. The selection of the combination of treatments and coatings as detailed in preceding paragraphs was the most compatible with the exposure medium.

4.6.5   Block I PSA Design

The Block I Power and Servo Assembly (PSA) contained all the electronics for the ISS except for a few modules located in the IMU and CDUs. The PSA also contained most of the electronics for the optical subsystem. It consisted of ten nickel-plated magnesium alloy trays, each containing a number of replaceable modules. There were a total of ninety modules in the ten trays. Each tray had a locating tongue that fitted into a groove in the PSA end connector assembly to insure proper alignment of electrical connectors. The tapered tongue also helped to maintain physical contact between the base of the tray and the thermal interface material which transmitted heat to the spacecraft coldplate. On the front of each tray was a test connector, guide pins to locate the tray and a captive fastening screw used to jack the PSA into place. A stiff metal strip, called a toe plate was fastened to the spacecraft structure and provided support for the front end of the trays. The fastening screws bolted into the toe plate. The modules plugged into feedthrough connectors on the vertical wall of the tray and were held in place with mounting screws.

Intra-tray wiring was done on the other side of the vertical wall. The wiring was encapsulated with polyurethane. Moisture protection of the test connectors was provided by individual gaskets and connector covers. Figure 4-35 shows a typical tray.

The PIPA and IRIG calibration modules contained selected components which were matched specifically to a single IMU. It therefore appeared advisable to package these electronics in a separate sealed PSA when the IMU was changed. This assembly was called the PIPA Electronics Assembly (PEA) in the command module and the Pulse Torque Electronics (PTE) in the LM. The Block II and LM electronics CDUs were identical except that two coarse align modules were not needed in the CM CDU and therefore were not plugged in.

4.6.6  Block II and LM Electronics

The Block II and LM electronics consisted of one CDU, one PEA or PTE, and one PSA. The PSA for the command module also contained the electronics for the optical subsystem. There were no OSS electronics in the LM PSA.

The Block II and LM PSAs, PEA, and PTE were all basically similar in package design. Each contained a machined header to which the modules were attached on one side, while the reverse side contained the inter-module wiring and output harnessing. A structural cover and gasket were bolted to the header on the module side to provide a hermetic seal. All electrical connections to these assemblies were through sealed connectors. Each assembly contained a slight positive pressure using an inert gas with a helium trace. The wiring was sealed by a solid polyurethane encapsulant.

The CDU for Block II and LM contained two separate header assemblies. The long modules mounted across the width and bolted into the header. The two headers were bolted together with modules back to back. A gasket between the

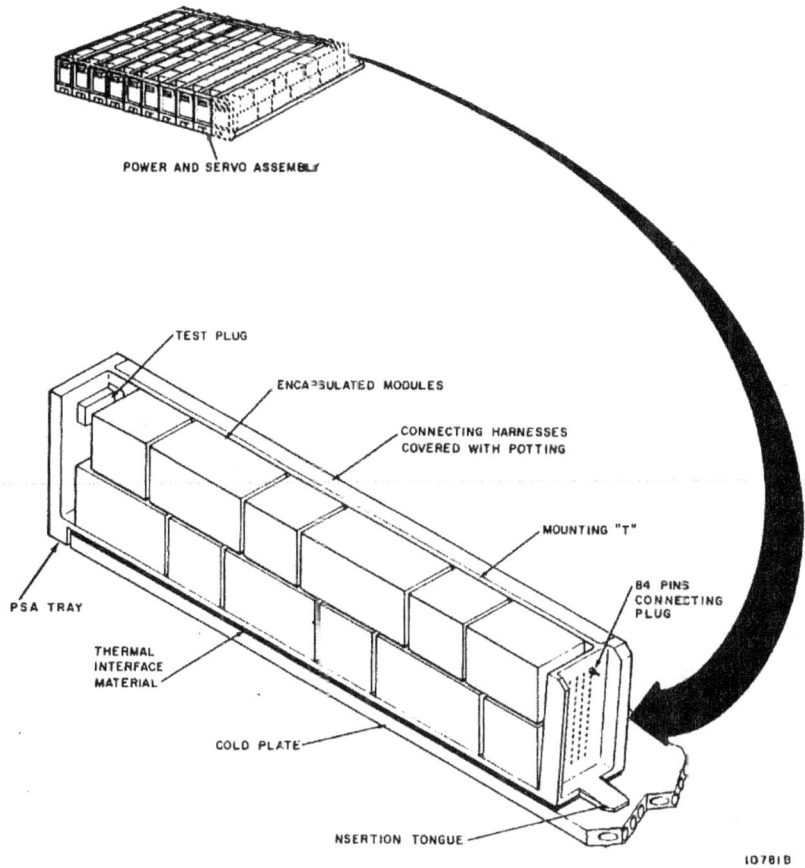

POWER AND SERVO ASSEMBLY

TEST PLUG

ENCAPSULATED MODULES

CONNECTING HARNESSES
COVERED WITH POTTING

MOUNTING "T"

84 PINS
CONNECTING
PLUG

PSA TRAY

THERMAL
INTERFACE
MATERIAL

COLD PLATE

NSERTION TONGUE

IO781B

Fig. 4-35  Power and Servo Assembly

headers provided a hermetic seal. In this configuration, the wiring planes were on the outside and sealing was provided by a solid polyurethane encapsulant. Light sheet metal covers provided mechanical protection for the encapsulation. The CDU was pressurized slightly above one atmosphere using an inert gas with a slight helium trace.

The gasket seals provided a near hermetic seal for the electronic packages and prevented leakage to levels below $10^{-4}$ standard cc/sec of helium across a one atmosphere pressure differential. A seal of this quality significantly limited the outgassing of toxic and malodorous materials from inside the sealed assembly and offered positive protection of the unit from moisture environment. The integrity of the seal could be 100% tested at the factory or in the field.

### 4.6.7  Module Design

Module design remained essentially similar between Block I and Block II. The module was    the lowest order replaceable unit. In order to obtain the utmost in reliability, the proven welded cordwood module was utilized. This configuration was used throughout the ISS in the Gimbal Mounted Electronics, the Power and Servo Assemblies, and the PIPA Electronics.

In lower heat dissipating circuits, the cordwood components were mounted between two epoxy fiberglass wafer boards and interconnected by welding. This assembly was then mounted in an aluminum (Block I) or magnesium (Block II) frame, welded to the output pins and foam-encapsulated.

Generally, any module over 10 Watts heat dissipation utilized an aluminum or magnesium heatsink. These were tape control machined to receive component parts and output connectors in cordwood fashion. Point-to-point welded interconnections were made on both sides of the module. After the module was tested electrically and inspected, the module was then encapsulated with foam potting.

The design of the CDU digital modules was dictated by the large number of integrated circuit flatpacks in this assembly. The CDU modules were long and narrow consisting of two fabricated interconnection matrices to which the flatpacks were bonded. The interconnection matrices were in turn bonded to a rib in the magnesium module header and connected to the output pins. Each matrix was built up of layers of etched metal conductors and Mylar insulators. The flatpacks were then welded to the conductors. CDU modules with analog components were of the same shape as above but with the heatsink machined to receive cordwood components.

Module geometry was derived by evaluating the following considerations: (1) function, (2) reliability, (3) repairability, and (4) producibility. Experience indicated that the functional circuit, as breadboarded and schematically drawn by the circuit designers, was very often the best compromise. Normally, package design affects the initial schematic only when the available volume is too small.

Depending upon the geometry of the module, two, three, or four screws in the frame held the module in place. These mounting screws were usually captivated screws that allowed the module to be mounted into place or dejacked. The CDU modules had two mounting and dejacking screws, one at either end.

Insertion of the module into the header connector was blind. Since there was some danger that the pins would be slightly bent or broken, hard metal alignment pins were used to guide the insertion of the module with proper connector pin alignment. When the module was not inserted in a header, a shielding skirt was provided to protect the male connector pins.

EMI bonding of all analog modules in the electronic packages was provided by a structural ground on the modules that consisted of a stainless steel pin mounted in a spotfaced-tapped hole in the module frame connected to an output pin.

4.6.8 Connector Design

The Apollo system connector was the Malco "Mini Wasp." It was a tuning fork friction connector with a nylon insulator and mounted on 0.125" centers. The electronic modules of all the packaged hardware were interconnected by means of Malco pins that were set into the module housing and mated with pins in the header structure. In general, the module was fitted with a male connector pin and the header or tray used the female pin. The male pin had a nickel base with a 0.000050 inch minimum gold-plating. This metallurgy was selected because it offered the best combination of desirable characteristics, namely: (1) corrosion resistance, (2) low surface resistance, and (3) compatibility with resistance welding. The female connector could be described as a wrappost with a beryllium-copper base, 0.000200 silver plate, and a final plate of 0.000050 minimum gold. The female wrappost was compatible with the mechanical wirewrap process.

The criteria used for evaluation of terminations, i.e., soldering welding, wirewrapping, or crimping were: (1) reliability, including joint life with respect to mechanical integrity and contact resistance, (2) compatibility with structure and its dynamic environment, (3) interconnection density required, (4) ease of repairability, and (5) time, skill and tooling required to make the termination.

In Block I, the module-to-tray connectors were environmentally sealed by individual module gaskets. When the inflight module replacement requirement was dropped, the environmental seal was supplemented by the application of Dow Corning DC-4 silicone grease around the connector pins. In Block II and LM the module-to-header interfaces were wholly within the sealed packages and required no additional environmental protection. During system installation into the spacecraft, all G&N external interface connectors had a liberal amount of DC-4 grease applied to the pins to aid in excluding moisture.

4.6.9   Thermal Design

4.6.9.1   General Considerations.   The thermal aspects of the electronics packaging were concerned with the removal of heat from the modules. This was accomplished by providing a thermal conductive path from the modules to spacecraft-supplied cold-plates. The coldplates were of an internal pin design sandwiched between two 0.020 inch thick plates. A water glycol solution, maintained between $32^{\circ}F$ and $55^{\circ}F$, was circulated through the coldplates. Typical thermal resistances from coldplates to coolant were on the order of $3.5^{\circ}F/Watt/in^2$.

4.6.9.2   Block I Thermal Design.   The Block I PSA design consisted of removable modules mounted to a vertical member of a removable tray. The requirement for inflight maintenance, with the requirement of handling the modules and trays, precluded the use of thermal conductive grease between the PSA trays, and the coldplate. Consequently, North American developed a thermal interface material consisting of a rubber-like tubing (1/8 inch OD) with a copper foil helically wound on the outside. This material was laid side by side to form a mat and was placed between the cold-plate and the PSA. Late in the Block I program, the inflight maintenance requirement was abandoned and a thermally conductive grease, Dow Corning DC-340, was used in conjunction with the thermal interface material to effect a better heat transfer.

Figure 4-36 shows a typical component mounting method and the resistances considered in the thermal analog. A nodal network was written for each tray, which included spreading within the tray down to the coolant. Solving this network gave the component temperature. For high component reliability, junction of silicon devices were not allowed to exceed $105^{\circ}C$. These measures insured control of the interface temperature at the tray base with NAA to not exceed certain temperature limits. These limits varied from $90^{\circ}F$ to $120^{\circ}F$ for the ten trays that comprised the Block I PSA.

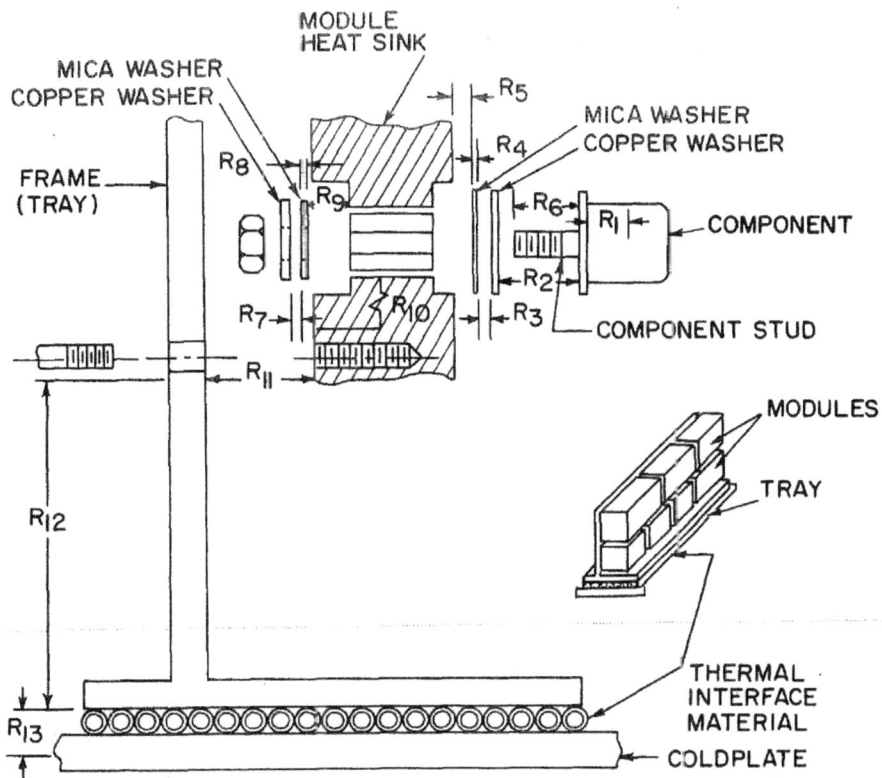

R1  JUNCTION TO CASE THERMAL RESISTANCE

R2-R9  SERIES/PARALLEL RESISTANCES THAT MAKE UP THE CASE TO HEAT SINK THERMAL RESISTANCE

R10  RESISTANCE THRU HEAT SINK

R11  HEAT SINK TO TRAY CONTACT THERMAL RESISTANCE

R12  THERMAL RESISTANCE THRU TRAY

R13  TRAY TO COLDPLATE THERMAL RESISTANCE

Fig. 4-36  Typical Component Mounting

4.6.9.3 <u>Block II Thermal Design.</u>  The Block II thermal design had the modules mounted directly to a header which in turn was mounted to the coldplate.  Figure 4-37 shows the typical packaging arrangement used and the heat flow paths from components in the modules to the coldplate.  Header contact with the coldplate was through thermal islands.  The cutaway material between islands was used for module-to-module interconnecting wire runs and power lines as well as for weight reduction. Sizing of the islands was based on a heat density not to have exceeded 3 Watt/in.$^2$ in order not to have excessive $\Delta$T's into the coolant from the header so that high coolant inlet temperatures could be used.

Module packaging is shown in Figure 4-38.  Two different modules were used: a metal heat sink type to package high power dissipating components, and an open frame type for low power dissipating components.  For equal volumes, the open frame type was considerably lighter, but the thermal resistance through foam was two orders of magnitude greater than that of a metal heat sink.

A thermal analog was assembled which coupled together the components, modules, header and cover to the coolant and surrounding environment.  Control of the interface temperature at the thermal islands insured that the silicon devices would not exceed reliability limits.  The temperature limit  for the thermal islands was  120$^{\circ}$F.

4.6.9.4 <u>CDU Thermal Design.</u>  The CDU heat flow paths are shown in Figure 4-39. This design had two tiers of logic and analog modules.  The heat flows from each module into the trays, down to the thermal rails and into the coldplate.

The logic  flatpacks were of a low power type such that thermal straps to transport away the heat within the modules were not necessary.  Analysis showed that the multilayer boards bonded to a metal frame were sufficient to remove the heat from the flatpack junction to the module/frame interface at a $\Delta$T of 18$^{\circ}$F.  The junction of the logic elements was not allowed to exceed 70$^{\circ}$C.

A thermal analog was assembled that coupled the modules and trays to the coldplate and local thermal environment.  The indications were that the thermal rail temperature was not to exceed 110$^{\circ}$F if the logic element junction of 158$^{\circ}$F (70$^{\circ}$C) was not to be exceeded.

It is important to note that the CDU was designed to be interchangeable between the LM  and CM  vehicles.

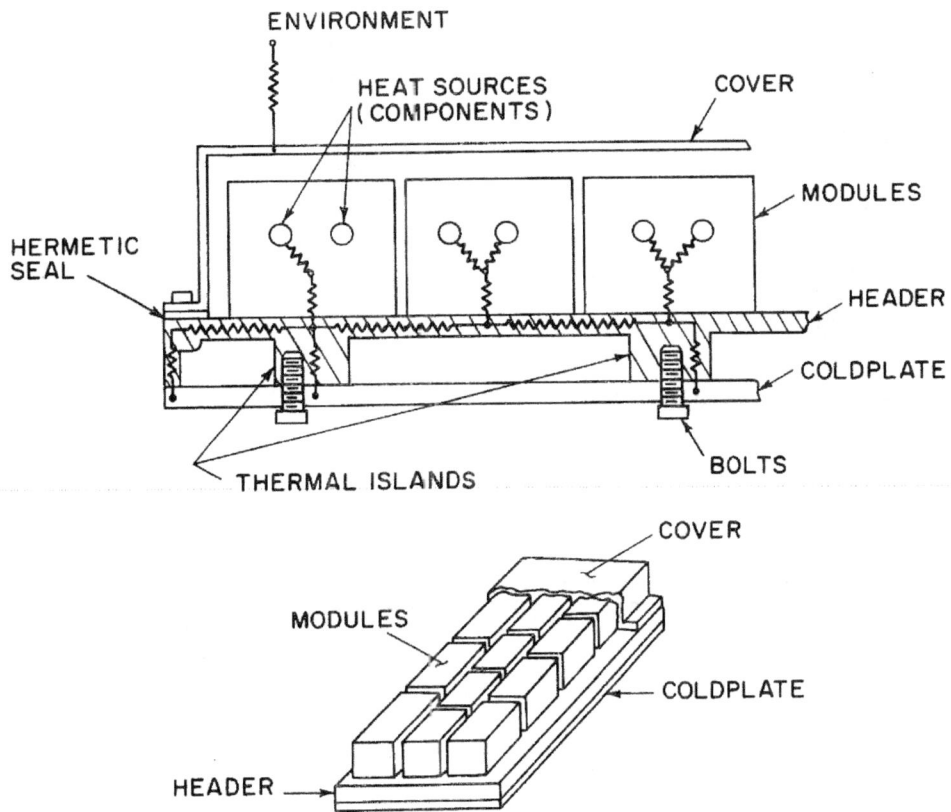

Fig. 4-37 Typical Block II Packaging

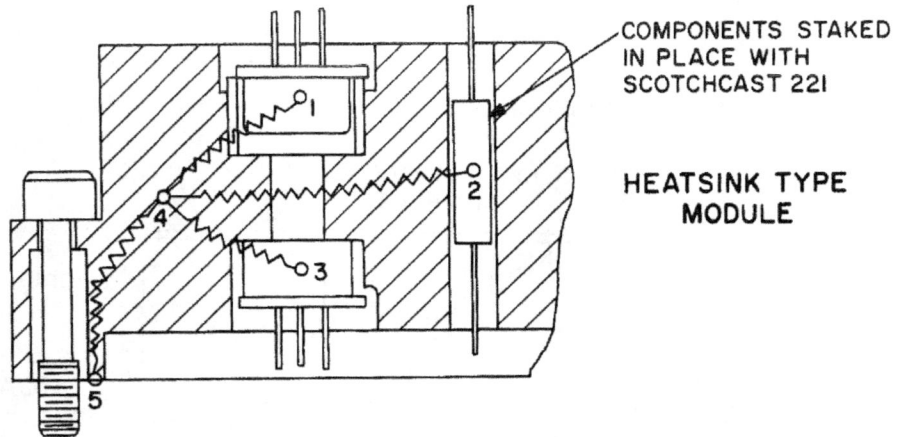

COMPONENTS STAKED
IN PLACE WITH
SCOTCHCAST 221

HEATSINK TYPE
MODULE

$R_{1-4}$, $R_{2-4}$, $R_{3-4}$ COMPONENT TO HEAT SINK RESISTANCE PATHS.

$R_{4-5}$ THERMAL RESISTANCE THRU HEATSINK.

MG FRAME

GLASS EPOXY
WIRING
BOARD

FRAME TYPE
MODULE

URETHANE
FOAM

ASS'Y NYLON SPACER

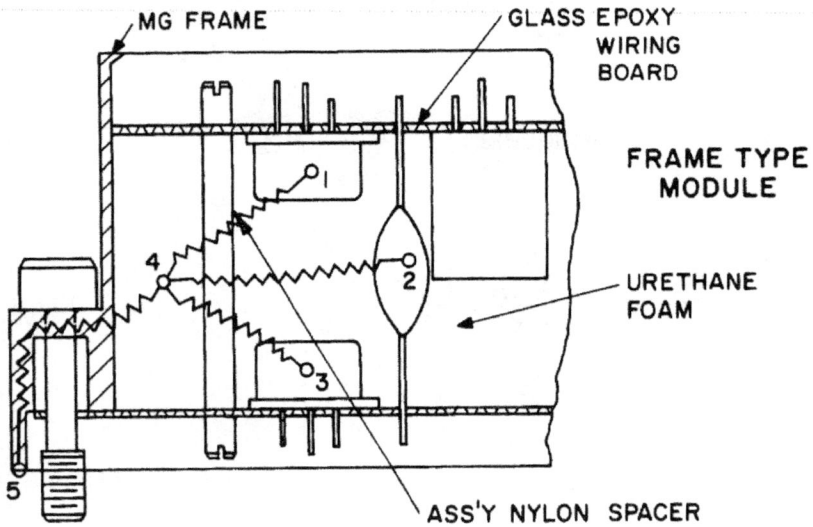

Fig. 4-38 Typical Module Packaging

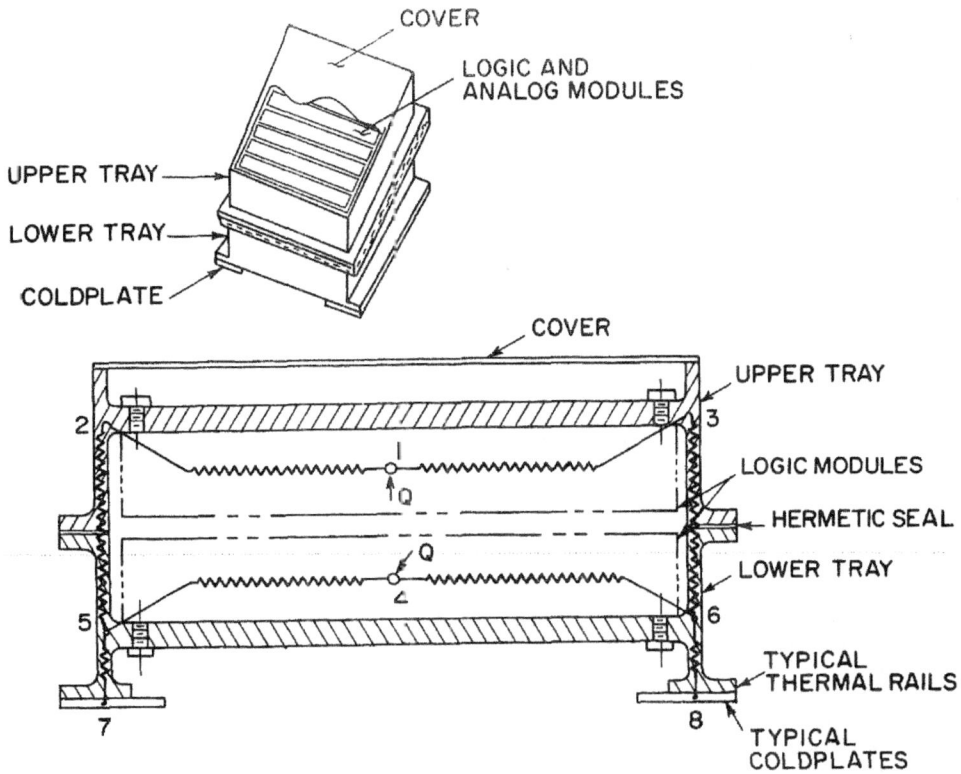

COVER

LOGIC AND ANALOG MODULES

UPPER TRAY

LOWER TRAY

COLDPLATE

COVER

UPPER TRAY

LOGIC MODULES

HERMETIC SEAL

LOWER TRAY

TYPICAL THERMAL RAILS

TYPICAL COLDPLATES

$R_{1-2}, R_{1-3}$   RESISTANCE FROM LOGIC ELEMENT JUNCTION
$R_{4-5}, R_{4-6}$   TO TRAYS

$R_{2-5}, R_{3-6}$   INTER-TRAY RESISTANCE

$R_{5-7}, R_{6-8}$   RESISTANCE FROM TRAY THERMAL RAILS TO COLDPLATE

Fig. 4-39  CDU Packaging

The automatic wirewrap process was utilized for the single plane header. Because of wire densities, the high reliability automatic interconnection scheme was suited to producibility and repairability. The lengths of the wires in each header caused some concern since they were a source of coupled noise where wires ran parallel for a finite length. The lengths and routing of the wires were determined by a computer program. The only restrictions placed on the program were those inherent in the wiring patterns available with the Gardner-Denver automatic machine.

4.6.10  Packaging Problem

(1)  A major mechanical difficulty in the Block I PSA was to achieve an adequate thermal interface between the PSA trays and the spacecraft coldplate. Tests of the thermal interface material showed that thermal conductance varied in direct proportion to the depth of its compression. Other tests indicated that the pressures required to deflect the material to achieve the desired conductance $(100 \, BTU/hr/{}^{o}F/ft^{2})$ were much higher than originally anticipated. These forces caused bowing at the trays and plate, thereby reducing conductivity across local areas on the interface. Closer tolerances for the tray locating tongues, stiffening of the trays, and changing the toe plate from aluminum to beryllium produced an adequate but marginal design. However, it was not until the inflight maintenance concept was dropped thus permitting the use of conductive grease on the thermal interface material, was the problem adequately solved.

(2)  The module designs were plagued by numerous modifications required by circuit and component changes. As was pointed out previously, scheduling constraints dictated a release to production concurrent with engineering evaluation testing. Nearly all the modifications resulted from circuit changes determined by this testing. Component changes were made to bring circuit parameters to optimum design centers, or because reliability determined that a particular component was unreliable. Wiring changes and component placement was altered to minimize electromagnetic coupling between circuits. Circuit changes were made when the original design was found to be marginal under adverse operating conditions. In some instances, high power dissipation components were relocated to remove local hot spots. Where possible, the changes were made as a "repair fix" by depotting or rework of manufactured modules, and the necessary changes being incorporated into forward production. Where changes were too extensive, modules were scrapped and replaced with new designs, the change from ternary to binary torquing of the PIPAs required new module design.

(3)  The Block I inflight module replacement feature required that the modules be removed using only a number 10 Allen wrench. The modules were fastened

to the trays with number 10 captive Allen head bolts, and necked down near the bolt head to provide clearance through a threaded portion of the module. There were numerous bolt failures in the early systems caused by shearing of the bolt heads. Necking down the bolts left an insufficient wall thickness in the region between the bolt shank and the Allen head recess. A bolt configuration change to increase the material thickness in that region plus a material change to a stronger bolt solved that problem.

(4) When the IMU gimbal mounted electronic packages were designated, it was felt that an added measure of quality control could be achieved if the modules were encapsulated in clear potting. It was reasoned that if one could see inside the module, greater care would be taken in the assembly of the module, and that fact would add to the reliability of the assembly. The cordwood assembly was packaged inside a transparent Lexan case then potted with a transparent potting material. The Lexan cases exhibited a high incidence of cracking and crazing. Numerous attempts to solve the crazing problem proved futile. Finally, a drawn aluminum case was designed for the gimbal mounted electronics, and the visual inspection feature was abandoned.

(5) A potential structural problem appeared during the environmental design evaluation of the Block II CDU. The response of the modules within the headers to vibration or shock inputs from the spacecraft structure was higher than anticipated. Although not a critical problem, the condition was not desirable. A corrugated metal damper plate was placed between the two arrays of modules to help restrain the modular response. This decreased the module resonance peaks to reasonable levels with a sufficient margin of safety.

(6) The CDU was fastened to the LM spacecraft with bolts that passed through clearance holes in the CDU flange, coldplate, and coldplate support structure and engaged nut plate fasteners on the far side of the support structure. The mounting bolts were chosen such that the bolt shank would pass completely through the clearance holes. Proper account was not taken of the imperfect lead in threads and resulted in bolts with too long a grip length. As a result, the bolts could not be screwed far enough into the fasteners to assure proper CDU mounting. This condition was discovered after the CDU was installed into the spacecraft. The bolts were then removed and replaced with correct size bolts. An investigation disclosed that a similar condition existed for the mounting bolts of the PSA, CDU and computer in both the Block II command module and the LM. Appropriate action was taken to correct the problem.

(7) Block I PSA modules utilized a block anodized aluminum housing and two types of encapsulation materials. The bottom end of the module was encapsulated with Stycast 1090, and the remainder of the module was encapsulated with polyurethane foam. After encapsulation, the bottom of the module was machined to obtain the required dimension from the bottom of the module to the bottom of the connector. An examination of several modules that failed during humidity qualification testing disclosed that the Stycast 1090 had separated from the housing and allowed moisture to penetrate the module. An engineering investigation determined that the adhesion of Stycast 1090 to black anodized aluminum was at best marginal. Silicone and hydrocarbon oils, acting as contaminates were found on the separated surfaces of one module. These contaminating agents were present in sufficient quantities to prevent adhesion. The forces imparted by the milling cutter during the machining operations were found to cause separation where low peel strength existed.

Satisfactory adhesion was obtained by first priming the aluminum housing with a thin coating of C7 epoxy adhesive. This change was incorporated in all forward production modules. In addition, measures were taken to insure that module components were free of contamination. The module machinery technique was revised to prevent the milling operation from imparting abnormal peel forces to the assembly.

Modules which had already been constructed, were repaired by mechanically removing a groove of Stycast 1090 around the periphery of the encapsulated region where it interfaced with the housing and filling it with C7 epoxy adhesive. There were no further module failures.

(8) Block I hardware developed shorts in the PSA tray header wire resulting from cold flow of the wire insulation. This phenomenon occurs when a Teflon-insulated wire is subjected to small but constant pressure against a sharp corner such as another Malco pin, mounting boss, or thermal island. The continuous pressure does not result in immediate cutting of the insulation but rather in gradual regression of the insulation. Proper selection of new wire insulations, such as Mylar and polyamides which are more resistant to cold flow and still compatible with the encapsulant material in the wirewrap plane, alleviated the problem.

Section V

SYSTEM FLIGHT EXPERIENCE

5.1   Synopsis

The Apollo GN & C system had successfully flown in seven flights as of 12 March 1969. This experience provided data for an identification of the elements of system design, prelaunch and flight activities that were most influential in achieving success.

The prelaunch, flight activities and data reviewed included four unmanned Apollo launches (three command modules and one lunar module) and three manned missions. Comparisons were made between ground measured data and measurements made during missions. The calculated system performance for some guidance phases of the mission were based upon ground measurements and compared to actual inflight performances and to system specified performance.

The review of the experience indicated that the significant factors enabling the Apollo GN & C system to successfully perform its function were the early recognition of necessary design changes for stable performance, the ability to predict the expected system performance, the discipline imposed by the policy of allowing no unexplained failures and the ability to diagnose flight operational anomalies.

The elapsed time of major items from the design inception to the first flight was less than five years.

Brief Time Schedule

| | |
|---|---|
| System Design Start at MIT | October 1961 |
| GN & C Installation in First Flight | |
| Spacecraft | September 25, 1965 |
| First Flight Program Release | January 1966 |
| First Flight | August 25, 1966 |

During this period of time, concepts of the lunar landing mission operations were changing and GN & C system requirements were added, subtracted, and modified. The system was designed to be fully integrated with the astronaut as well as to have an automatic capability. The first four flights were unmanned and required the automatic system. The original design intent was to have a completely self-contained navigation system. During the program, it was stipulated that primary navigation would be the function of the ground-based tracking network. Both means of navigation were accomodated as ground-transmitted spacecraft state vectors.

5.2    Prelaunch Operation

The Apollo GN & C system on the launch pad at KSC was subjected to approximately 12 months of system testing. Then a final test for verification of flight readiness was performed. When the flight readiness test was successfully completed, the GN & C system was ready for the mission. The countdown operations followed.

The average lunar module GN & C system was checked out for several weeks prior to the scheduled flight. The computer erasable memory was then loaded for flight operations and the system turned off, except for IMU temperature control. The system was not activated again until it was in space. The average CM GN & C system was operated fifty hours in support of the countdown. The system was exercised through automatic operational checks and a final calibration test. The initial conditions for the mission were loaded into the computer erasable memory and, by gyro compassing, the inertial measurement unit was commanded to start the automatic platform alignment. About two weeks prior to launch, the alignment of the inertial measurement unit was verified by the astronaut using the optical system space sextant to sight on illuminated targets two miles from the launch vehicle. The launch vehicle had demonstrated enough stability so that optical verification was not required in the final countdown.

The control room for the spacecraft checkout and launch was located twelve miles from the launch site in the MSOB (Manned Spacecraft Operations Building Figure 5-1). In the control room, the serial digital data from the spacecraft was processed by the ACE (Acceptance Checkout Equipment) computers which in turn displayed the information to the test engineers as meter and oscillograph readings, event lights or CRT (cathode ray tube) displays. In addition to standard data the telemetry transmitted from the flight computer to the ground was processed to produce a CRT display analogous to the onboard DSKY display that the astronauts were monitoring. The K-START (Keyboard Sequence to Activate Random Testing) command system duplicated the keyboard section of the onboard computer DSKY. The keyboard entry was paralleled with a tape reader allowing for automatic, rapid, error-free command sequences from the control room to the onboard computer.

The capability for remotely monitoring and commanding the GN & C system was exploited in the design of the prelaunch test procedures to enable paralleled testing of spacecraft subsystems.

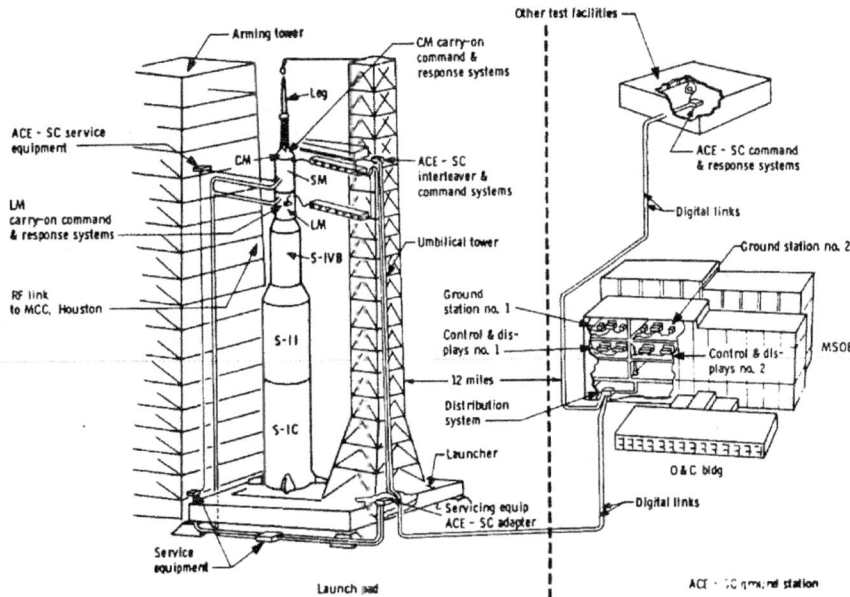

Fig. 5-1 Apollo Prelaunch Operations

## 5.2.1  Prelaunch Checkout Design Objectives

The Apollo guidance computer was programmed to compensate the system for the predominant instrument errors.  The objective of the prelaunch calibration testing was to provide test estimates of the present values of the error coefficients for use as compensation and to provide data for determining the uncertainties to be expected.

The unique characteristics of an inertial system utilizing a general purpose digital computer with a remote control capability were exploited in the design of the prelaunch calibration tests. The guidance system calibration test requirements were designed to minimize the launch preparation time. The test method utilized gravity to eliminate the need for external references. The known amplitude of gravity was used to calibrate the accelerometers. The gyro drift calibration was based on the detection of the vector rotation of gravity by the accelerometers. The drift information had to be separated from accelerations caused by launch vehicle acceleration resulting from sway and from noise as a result of quantization in the pulsed integrating pendulous accelerometer. The velocity quanta size for the CM was 5.85 cm/s and for the LM, 1 cm/s. The information was separated from the noise by a simplified optimum linear filter, which included in its state vector, estimates of launch vehicle disturbances.

The measurements made on the launch pad were usually used as reconfirmations of the selected compensation values. The compensation parameters were accelerometer bias and scale-factor errors for the three accelerometers, and gyro bias drift and two acceleration-sensitive drift terms for the three gyros, for a total of fifteen terms.

5.2.2  Error Analysis for Prelaunch System Flight Worthiness Demonstration

The measurements made prior to launch were used as indications of uncertainties to be expected during a mission. The prelaunch system performance data had specified tolerances. In the cases where the specified tolerances were exceeded, the flight worthiness of the system was evaluated on the basis of the probable mission effect of the deviating parameter. As an example, shifts of gyro drift parameters beyond specified limits during prelaunch tests occurred on Apollo 3, 4, 5, 6. Decisions about the flight worthiness of those systems were made by first classifying the problem as indicating possible catastrophic failure in flight or one indicating performance degradation. In cases where reliability problems were suspected, the Inertial Measurements Unit was replaced ( Apollo 6 ). In the other cases, where the test data showed a performance degradation, determination of the mission effect was required. This determination required the development of error analyses that related variations of each of the measurable parameters to the mission.

Each mission in the Apollo program is unique. A separate error analysis is to be performed for each. The mission performance requirements, based upon a typical lunar landing, were defined early in the Apollo program. Because of the variety of missions and mission objectives, it was necessary to have a separate error analysis for each mission. For all missions except Apollo 5, the segmented mission phase approach to error analysis using a linearization technique was entirely adequate and was pursued. An error analysis was conducted using both the specification values, as well as the demonstrated values, for the GN&C system. A comparison of specification, actual ground measurement, and flight results for selected mission phases is presented in the following table and Figures 5-2 through 5-6.

| Mission and Parameters | 1-σ Uncertainty Based on | | Best Estimate Error |
| --- | --- | --- | --- |
| | Specified Performance | Actual Pre-Flight Data | |
| **Apollo 4 (SA501)** | | | |
| 1. Position error at re-entry start | 2.75 nm | 3.15 nm | 7.5 nm* |
| 2. Velocity error at re-entry start | 26.6 ft/s | 51.5 ft/s | 140 ft/s* |
| 3. Position error at splash | 22.5 nm | 18.6 nm | 7.4 nm* |
| Note: *NASA-5-68-454 | | | |
| **Apollo 5 (LM1)** | | | |
| 1. Altitude uncertainty at perigee after APS cutoff | 100,890.2 ft | 109,079.7 ft | Unavailable |
| 2. Position error indicated at SIVB cutoff | 5.6 nm | 4.22 nm | 0.0 nm |
| 3. Velocity error indicated at SIVB cutoff | 132.5 ft/s | 100 ft/s | 2 ft/s |
| **Apollo 6 (AS502)** | | | |
| 1. Position error at re-entry start | 2.8 nm | 2.75 nm | 2.7nm* |
| 2. Velocity error at re-entry start | 58 ft/s | 57 ft/s | 10.2 ft/s* |
| 3. Position error at re-entry end | 14.2 nm | 7.2 nm | ** |
| Notes:<br>*MSC-PA-R-68-9<br>**Due to failure of the SIVB to re-ignite, the re-entry trajectory was not as planned; therefore, the entry error is meaningless. | | | |
| **Apollo 7 (AS205)** | | | |
| 1. EOI cutoff position uncertainty | 3.1 nm | 1.8 nm | 2.6 nm |
| 2. EOI cutoff velocity uncertainty | 73 ft/s | 43 ft/s | 60 ft/s |
| 3. Rendezvous TPI burn position uncertainty | 1.95 nm | 0.7 nm | 0.51 nm |
| 4. Rendezvous TPI burn velocity uncertainty | 13.7 ft/s | 5 ft/s | Unavailable |
| 5. Position uncertainty at drogue deploy | 2.8 nm | 1.4 nm | 2.2 nm |
| 6. Velocity uncertainty at drogue deploy | 56 ft/s | 33.7 ft/s | Unavailable |
| **Apollo 8 (AS503)** | | | |
| 1. EOI cutoff position uncertainty | 4.3 nm | 3.9 nm | 0.016 nm |
| 2. EOI cutoff velocity uncertainty | 70.7 ft/s | 66 ft/s | 1 ft/s |
| 3. TLI cutoff position uncertainty | 1.25 nm | 1.1 nm | 1.9 nm |
| 4. TLI cutoff velocity uncertainty | 12.2 ft/s | 10 ft/s | 18 ft/s |
| 5. Perilune uncertainty following LOI (3 ) | 0.31 nm | 0.23 nm | 0.15 nm |
| 6. Apolune uncertainty following LOI (3 ) | 4.7 nm | 2.2 nm | 1.46 nm |
| 7. Position uncertainty at drogue deploy (CEP) | 1.92 nm | 0.96 nm | 0.815 nm |

TABLE 5-1

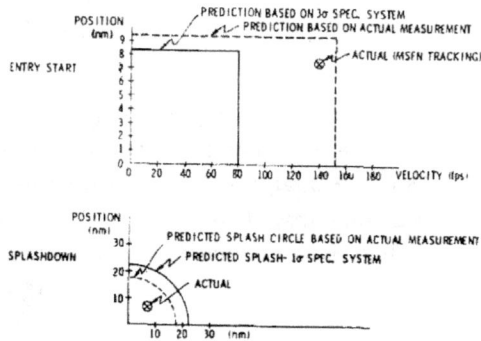

Fig. 5-2 Comparison of Predicted and Actual Entry
Errors on Apollo 4

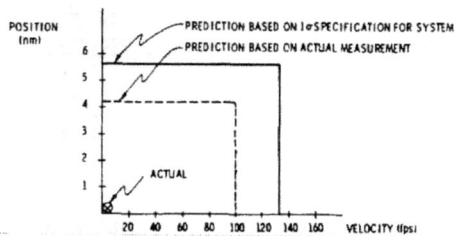

Fig. 5-3 Comparison of Predicted and Actual SIVB Cutoff
Errors on Apollo 5

The unmanned Apollo 5 flight was such that known initial conditions for each
thrusting phase were not available. As the system guided the vehicle based upon
its actual set of initial conditions, the guidance errors could not be treated with
linearized perturbations. The resulting position and velocity errors became more
nonlinear as the mission progressed. The mission was scheduled for nine earth
orbits, and the small-angle assumptions usually used with gyro drift were no longer
applicable. The only solution was to conduct a large number of Monte Carlo error
analyses of the complete mission.

Some interesting examples of how error analysis helped resolve operational
problems that occurred on the early flights are included here for the reader's evaluation. The flight plan for AS-202 called for a sub-orbital flight of approximately 3/4
of an orbit with a maximum entry range coupled with a maximum heat-rate input to
the heat shield. The original requirements called for an entry-angle uncertainty
specification of $1/2^\circ$. This was an easy achievement with the ground giving a state-
vector update. During the checkout phases of the vehicle, it was learned that there
were phases in the mission program when an update should not be sent

ENTRY START
POSITION
(n mi)

5

4

3

2

1

ACTUAL (BASED ON MSFN)

PREDICTION BASED ON 1σ SPECIFICATION FOR SYSTEM

PREDICTION BASED ON ACTUAL MEASUREMENT

10  20  30  40  50  60     VELOCITY (fps)

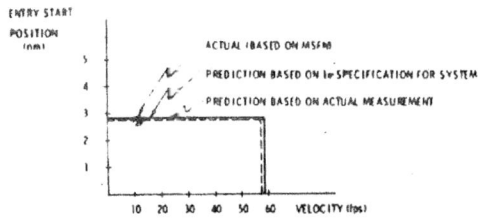

Fig. 5-4  Comparison of Predicted and Actual Apollo 6
Entry Errors

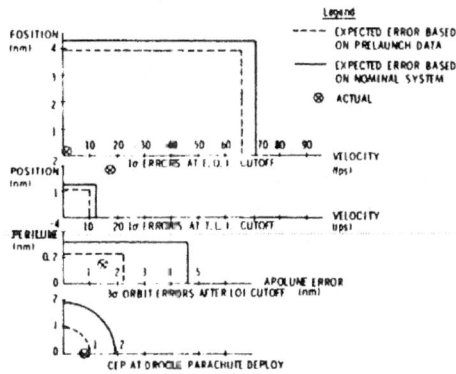

Legend

---- EXPECTED ERROR BASED
ON PRELAUNCH DATA

—— EXPECTED ERROR BASED
ON NOMINAL SYSTEM

⊗  ACTUAL

POSITION
(n mi)    4

3

2

1

10  20  30  40  50  60    70  80  90    VELOCITY
(fps)

1σ ERRORS AT T.L.I. CUTOFF

POSITION
(n mi)
1

VELOCITY
(fps)

10    20  1σ ERRORS AT T.L.I. CUTOFF

PERILUNE
(n mi)

0.7

0

1    2    3    5    APOLUNE ERROR (n mi)

1σ ORBIT ERRORS AFTER LOI CUTOFF

2

1

0    1    2

CEP AT DROGUE PARACHUTE DEPLOY

Fig. 5-5  Comparison of Apollo 8 Predicted and Actual
System Errors

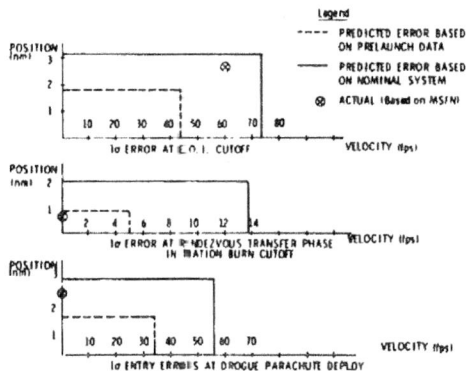

Legend

- - - - PREDICTED ERROR BASED
ON PRELAUNCH DATA

—— PREDICTED ERROR BASED
ON NOMINAL SYSTEM

⊗  ACTUAL (Based on MSFN)

POSITION
(n mi)  3

2

1

⊗

10  20  30  40  50  60  70  80
1σ ERROR AT E.O.I. CUTOFF       VELOCITY (fps)

POSITION
(n mi)  2

1

2  4  6  8  10  12  14
1σ ERROR AT RENDEZVOUS TRANSFER PHASE     VELOCITY (fps)
INITIATION BURN CUTOFF

POSITION
(n mi)

2

1

10  20  30  40  50  60  70
1σ ENTRY ERRORS AT DROGUE PARACHUTE DEPLOY     VELOCITY (fps)

Fig. 5-6  Comparison of Apollo 7 Predicted and Actual
System Errors

because of onboard software deficiencies. This resulted in a condition where a backup system would be required for guidance. As checkout proceeded, it was clear that inertial performance could, with a $3\sigma$ uncertainty, not exceed $1/3^{\circ}$. However, near the flight readiness test, the performance requirement was stated to be $0.05^{\circ}$ with a $3\sigma$ uncertainty. The system would not have made it without update and might not have even with update. However, near launch the requirement of $1/2^{\circ}$ was reimposed and no update was attempted. Post-flight analysis showed the entry angle error to be $0.12^{\circ}$.

Another operational consideration where the error analysis was used concerned notification to the GN & C system that launch vehicle lift-off had occurred. This discrete command given to the spacecraft guidance computer was to change the mode of operation from gyrocompass to boost monitor. Three methods were used to achieve this: (1) At a time about five seconds before lift-off, a discrete command called Guidance Reference Release (GRR) was given; (2) At lift-off, the same hardwire discrete that went to the launch vehicle guidance system was also sent to the Apollo GN & C system when the vehicle actually lifted off; and (3) A backup lift-off command could have been sent to the computer either by the Astronaut or by an uplink command from the Mission Control Center at Houston.

At T-15 seconds, monitored and progressed by a digital computer, the Saturn vehicle countdown proceeds automatically. Holds had occurred after T-5 seconds and it was common practice to recycle back to T-15 minutes thus creating a possible problem. Should a sequence like this occur, the guidance system would be released and proceed to monitor the boost. Should recycle occur, there would be insufficient settling time to re-establish orientation of the GN & C system by gyrocompassing. The error analysis results indicated that the GN & C system would navigate and monitor boost properly even if it were released well ahead of lift-off. Because of program considerations, it was decided to remove the GRR signal and to launch with only two methods of indicating lift-off.

5.2.3  Checkout History and Experience
The Apollo system spends a majority of its life in checkout. The following table summarizes the history of systems to date. The average number of operating hours accumulated in checkout was 2460 hours during an average 10.45-month spacecraft testing period.

| System | Spacecraft Contractor's Plant | | | Kennedy Space Center | | G&N System | |
|---|---|---|---|---|---|---|---|
| | Installation Completed | System Removed/ Reinstalled | Shipped to KSC | System Removed/ Reinstalled | Launch Date | Months in Spacecraft | Operation Hours |
| Apollo 3 AS202 G&N 17 | 1/ 6/66 | None | 4/16/66 | None | 8/25/66 | 8.7 | 2192 |
| Apollo 4 AS501 G&N122 | 8/29/66 | None | 12/22/66 | None | 11/ 9/67 | 14.3 | 2907 |
| Apollo 5 LM1 G&N603 | 11/12/66 | IMU replaced 12/66 | 6/23/67 | Replaced Computer 6/67 IMU 7/67 | 1/22/68 | 14.3 | 2626 |
| Apollo 6 AS502 G&N123 | 1/ 3/67 | 6/67 | 11/23/67 | Replaced IMU | 4/ 4/68 | 8.6 | 2669 |
| Apollo 7 AS205 G&N204 | 12/16/67 | None | 5/30/68 | None | 10/11/68 | 10.0 | 2345 |
| Apollo 8 AS503 G&N208 | 4/ 1/68 | None | 8/12/68 | None | 12/21/68 | 8.6 | 1905 |
| Apollo 9 AS504 CM104 G&N209 | 5/ 2/68 | Replaced DSKY | 10/ 5/68 | Replaced IMU | 3/ 3/69 | 10.0 | Unavailable |
| LM3 G&N605 | 10/ 7/68 | None | 6/14/68 | Replaced IMU twice | | 17.0 | Unavailable |

TABLE 5-2

The success in meeting schedules and establishing the flight worthiness of all the hardware was attributed to early recognition of the importance of considering checkout problems in the design, to minimization of equipment removals by carefully reviewing all anomalies for flight impact, and to the discipline imposed by allowing no unexplained failures.

Spacecraft testing revealed that there was a high probability of applying and/ or removing spacecraft power to the GN&C system in an incorrect sequence. The first system design did not incorporate protective features for making the system tolerant of incorrect power sequencing. Incorrect power sequencing resulted in performance shifts of inertial components. Consequently, the design was changed to provide internal protection against such an occurrence which, subsequently, saved many hours of retest and stabilized the performance data obtained in spacecraft testing. Another example involved ground potential changes in docked test configuration. The possibility of reverse potential on the system was not considered in the initial design. When spacecraft tests indicated that reverse voltages could exist as a result of grounding configurations, the GN & C system electronics design was changed to tolerate reverse voltages.

The prelaunch checkout had to guarantee that the equipment would operate during the mission. When any discrepancy existed, positive action was taken to eliminate possibility of failure in flight. An example of this was the failure of the GN & C system to accept an entry mode change command once during checkout of the AS-202 system. Even though the problem was never duplicated, the relays that could have caused this single malfunction were replaced. Another example involved the computer in the same mission. While one of the computers was undergoing inspection at the factory, it was discovered that one of the vibration isolation pads was missing from the oscillator module. Subsequent examination of other available modules revealed that, on the basis of the sample examined, there was about a 20% chance that one of the vibration isolation pads was missing in the computer in the spacecraft. The decision taken 30 days prior to flight was to remove the computer and inspect. It was rapidly done and verified that the pad had been installed.

The early GN & C system operations were plagued by the occurrence of unexplained restart[*]. The concept of NO unexplained failures required that each restart be explained. The computer restarts were frequent early in the program but as effort was applied to explain each one they were reduced to zero. Noise susceptibility in test connectors was discovered and corrected by a computer shorting plug. Software errors were discovered and corrected by new software. Procedural errors were discovered by means of ACE playbacks and laboratory verification. The solution therefore involved hardware changes, software changes, procedural changes and, above all, education and understanding on the part of all GN & C system operation personnel. The successful operation of the hardware during the Apollo flights was, primarily, the result of this careful disciplined engineering that examined all facets of the situation and left no area uncorrected.

## 5.3    Flight Operations

During a mission the GN & C operation was monitored by computers in the Real Time Control Center (RTCC) in Houston. The digital data generated by the

---

[*]A restart is an internal protective mechanism that enables the computer to recover from random program errors, operator errors, and from environmental disturbances. Restart attempts to prevent the loss of any operating functions.

onboard computer consisted of lists of two hundred 14-bit computer words transmitted once every two seconds. The contents of the lists were designed to provide information relevant to the mission activity. The data were used to drive displays on the guidance officer's console and numerous other support consoles. The amount of data from the guidance computer was limited by the word size and transmission rate. The program design, however, allowed the program to select the quantities to be transmitted and thus compensated for this deficiency. The data used for the real-time displays were selected prior to the mission and were based on the flight controller's experience and operational requirements. In real time, the data format was quite inflexible.

The control of the system was accomplished in the same computer complex. The data transmission paralleled the onboard keyboard-entry capability. The data transmitted consisted for the most part of an update of the spacecraft position and velocity, which was determined by ground tracking stations and converted into the proper format by the Houston RTCC. The controller had the capability of commanding the spacecraft computer through an analogous keyboard with the same codes as those of the astronauts.

Review of the data obtained from flight monitoring indicated that the ground calibration enabled accurate error compensation. Review of the anomalies in flight operations indicated that there was a reasonable amount of time available during the mission for troubleshooting and diagnosis of problems. The only cases that could not be diagnosed in real time involved inadequate real-time data.

5.3.1 Guidance System Monitoring During a Mission

The monitoring of the guidance system performance during the mission consisted of comparing navigation data from other sources (ground tracking, Saturn V guidance, LM backup for CM, CM backup for LM), computing accelerometer output with no input at zero gravity, and determination of the quality of the inertial reference by successive inflight optical realignments of the IMU. These successive realignments were performed several hours apart so that the rotations of the IMU stable member required to realign it were mostly the result of gyro drift with the fixed errors reduced inversely proportional to this time interval. There were also operational techniques utilizing star and planet horizons for checking the commanded attitude prior to a velocity-change maneuver.

The onboard measurement of the available IMU performance parameters could have been used to further improve the performance. The compensation parameters could be modified through the keyboard, either onboard or from the guidance officer's console in Houston.

The guidance system monitoring was designed to provide the flight controllers with data upon which a prediction of the future operation of the system could be made. The flight controllers had pre-programmed decision points that enabled the mission to continue with a backup system in control, or with a new mission plan if their data indicated that the primary system might not perform adequately during the ensuing critical mission phase.

The data telemetry from the spacecraft was limited; hence, the ability to predict future operation was difficult. The limits set for the various parameters were selected on the basis of the worst performance experienced during design evaluation tests and prelaunch tests, excluding catastrophic failures.

Onboard measurements to date have indicated excellent performance. The only onboard measurement available for the unmanned missions (Apollo 4, 5, 6) was accelerometer output at zero gravity ($a_b$). The manned missions also included inertial platform drift at zero gravity (NBD). The inertial component data are presented in the following table and in Figure 5-7.

Figure 5-8 shows the monitoring of the IMU over a long period of continuous operation. The data indicate that stability of inertial operation was achieved in the design. The entire component data history is presented in Figures 5-8 and 5-9.

5.3.2 Mission Diagnosis

The adequacy of all subsystems to continue into the next phase and to complete the mission was reviewed continuously by the flight controllers. It was important therefore to diagnose problems in real time in support of the GO/ NO-GO decisions. The flight experience showed that there was adequate time available for problem diagnosis and that there was a capability for real-time troubleshooting. There were two types of problems where real-time troubleshooting was of no value. These were problems involving actual hardware failures and involving incompatibilities resulting from inaccurate models of the spacecraft used in the control programs.

| | Accelerometer Bias | | | Gyro Bias Drift Less Compensation | | |
|---|---|---|---|---|---|---|
| | $Ab_x$ $(cm/s^2)$ | $Ab_y$ $(cm/s^2)$ | $Ab_z$ $(cm/s^2)$ | NBDX (meru)[1] | NBDY (meru) | NBDZ (meru) |
| **Apollo 4** | | | | | | |
| In-flight measurement | 0.304 | 0.23 | -0.39 | | | |
| Compensation | 0.41 | 0.21 | -0.28 | | | |
| **Apollo 5** | | | | | | |
| In-flight measurement | 0.1 | -0.35 | 0.0 | | | |
| Compensation | 0.14 | -0.22 | 0.12 | | | |
| **Apollo 6** | | | | | | |
| In-flight measurement | -0.83 | 2.77 | 1.93 | | | |
| Compensation | 0.64 | 2.9 | 2.1 | | | |
| **Apollo 7** | | | | | | |
| During the Apollo 7 mission the crew removed power from the guidance system during inactive periods. Data was gathered on gyro drift and accelerometer bias. | | | | | | |
| Last prelaunch measurement | 0.2 | 0.24 | 0.16 | 1.9 | 0.4 | -0.8 |
| a. Accelerometer and gyro data following boost. In-flight measurement | 0.275 | 0.0 | 0.215 | 2.2 | 0.2 | 0.15 |
| b. Accelerometer and gyro data at 145 hours following several on-board removals and re-applications. In-flight measurement | 0.309 | 0.0 | 0.208 | 1.4 | -0.63 | 0.0 |
| **Apollo 8** | | | | | | |
| The Apollo 8 mission was flown with the guidance system continuously operating. The monitoring of the inertial reference and accelerometer errors provides us with a large set of data on Apollo inertial system performance in space environment. | | | | | | |
| Expected value from last ground measurement | 0.0 | 0.845 | 0.615 | 0.93 | 2.2 | 1.3 |
| a. Accelerometer and gyro data following boost In-flight measurement | 0.0 | 0.83 | 0.62 | 1.5 | 0.62 | 1.8 |
| b. Accelerometer and gyro data during translunar coast In-flight measurement | 0.0 | 0.83 | 0.605 | 1.51 | -0.13 | 1.84 |
| c. Accelerometer and gyro data in lunar orbit In-flight measurement | 0.0 | 0.83 | 0.60 | 1.6 | 0.03 | 1.97 |
| d. Accelerometer and gyro data during transearth coast In-flight measurement | 0.0 | 0.82 | 0.59 | 1.38 | 0.16 | 1.6 |
| **Apollo 9** | | | | | | |
| Expected value based on ground measurements | 0.38 | -0.004 | 0.002 | -1.6 | -0.4 | 2.7 |
| a. LM system after turn-on in orbit In-flight measurement | 0.32 | 0.013 | -0.008 | -3.6 | -0.1 | 3.3 |
| b. CM system after turn-on in orbit In-flight measurement | -0.53 | -0.34 | 0.38 | -2.3 | -0.5 | -1.6 |
| Expected value based on ground measurements[2] | 0.64* | -0.10* | 0.36 | -1.2 | -0.2 | -2.4 |

Notes:
(1) One meru is 0.015 degree per hour.
(2) The compensation value was changed in orbit.

TABLE 5-3

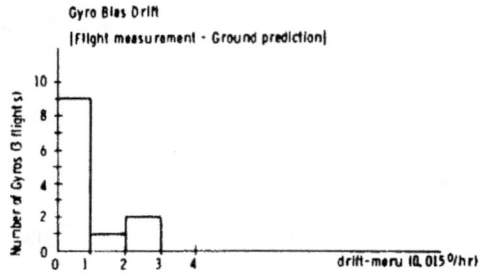

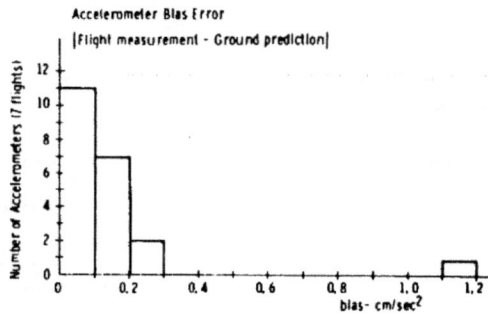

Fig. 5-7  Gyro Bias Drift (NBD) and Accelerometer Bias
Flight Data

GYRO DRIFT

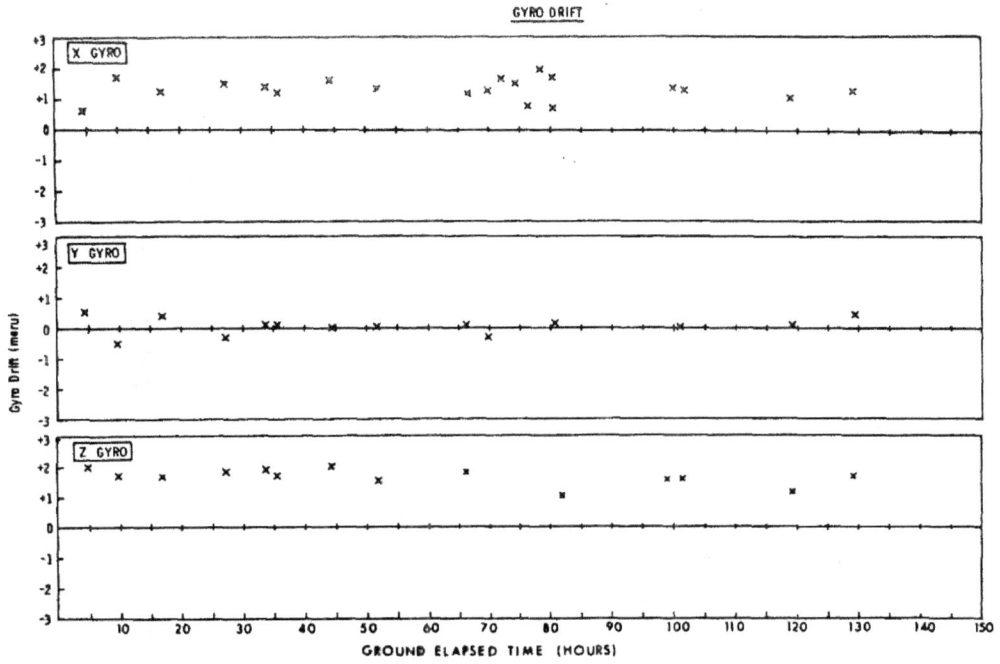

Fig. 5-8  Gyro Drift

PIP BIAS

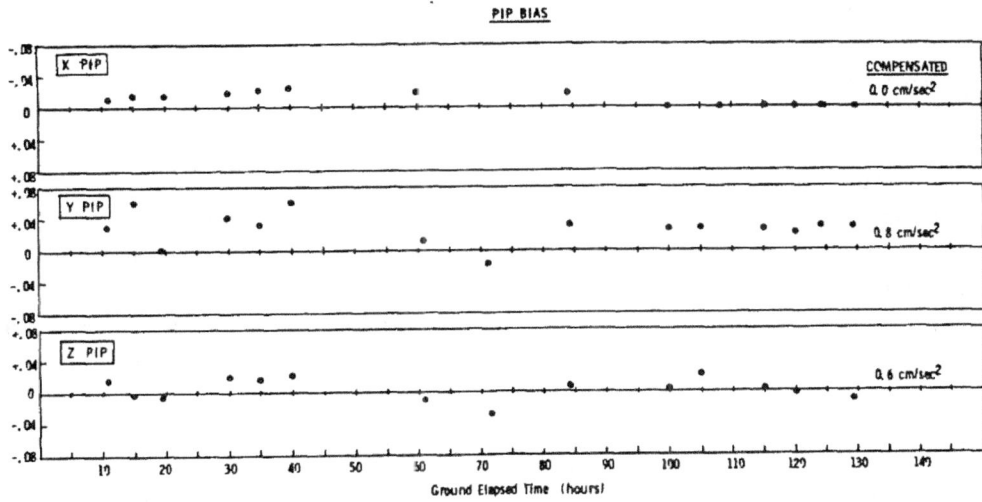

Fig. 5-9  PIP Bias

The following examples of problems involving the GN & C system that have been explained in real time illustrate the capability that does exist:

(1) APOLLO 4 (AS501). During the mission, it was reported that a large difference existed between the $\gamma$ indicated by the onboard computer and the $\gamma$ as compared from radar tracking data. $\gamma$ is the angle between the position vector and the velocity vector. Real-time measurement of accelerometers indicated the GN & C system was operating properly. The difference was found to be a ground computation error. The guidance system was allowed to continue in control of the mission.

(2) APOLLO 6 (AS502). During the mission, a divergence was observed between the attitude information supplied by the GN & C inertial reference and the backup body-mounted attitude gyros. The divergence was first attributed to GN & C malfunction. Real-time review of prelaunch data for the backup system indicated that the drift rates measured on the ground accounted for the divergence. The GN & C system remained in primary control for a successful mission.

(3) APOLLO 7 During the mission, a procedure for using the landmark-tracking navigation program for navigation sightings on the horizon was determined. The procedure, however, did not work in the spacecraft. The computer was programmed with a reasonable assumption that landmarks would be on the surface of the earth. The attempt to use the program for horizon sightings above the surface of the earth rather than the landmarks resulted in the attempt to compute the square root of a negative number. This resulted in a restart. The error in the procedure was quickly determined by ground tests.

a. Computer Restart--The Apollo computer had a catalog of navigation stars identified by numbers. The astronaut, by keying in a numerical code, told the computer the star to be used. The restart was the result of the astronaut not selecting any star when the computer requested a star selection. The computer interpreted the selection of "no" star as star number 0; the catalog, however, started with star number 1. The result was a computer restart resulting from accessing a memory location address that did not "exist." The restart was diagnosed from real-time displays.

b. Mark Button "Failure"--The computer assimilated line-of-sight data from the optics only upon astronaut command, which consisted of an interrupt caused by depressing the "mark" button on the navigator's control panel. The line-of-sight information was used for rendezvous navigation as well as for inertial platform realignment. To protect the rendezvous navigation information from being modified by platform alignment sighting data, the computer programmers prevented the processing of alignment "marks" during rendezvous navigation. The problem occurred when the Astronaut terminated the rendezvous navigation program in a fashion not expected by the programmers. This termination left the computer with the information that no alignment "marks" were to be processed. The next attempt at realignment failed as a result of an apparent failure in the "mark" interface to the computer. Ground troubleshooting uncovered the cause and a reselection and proper termination of the navigation program eliminated the problem.

c. Accelerometer Bias Change--The accelerometers in the Apollo inertial measurement unit were Pulsed Integrating Pendulous Accelerometers, PIPAs. The accelerometer used a pendulous mass as a torque summing element. The accelerometer bias (output with no input) resulted from the residual torques in the instrument. During a mission, at zero gravity, the accelerometer was calibrated by monitoring its output. During Apollo 7, the flight controllers noticed that the expected low output at zero gravity decreased to zero. This was interpreted as a possible hardware failure and an inflight test was conducted to determine if the instrument was operating properly. The test consisted of a maneuver to thrust along both directions of the accelerometer input axes. The results showed that the instrument was operating properly. The cause for the lack of any output was simply the PIPA reaching an operating region in free-fall where the torque generated by electronic nonlinearities was equal and opposite to the residual electromagnetic torques, and this yielded zero bias.

(4) APOLLO 8 (Prelaunch Alignment during the Transearth Coast) The commanding of the Apollo guidance computer consisted mainly in selecting numerically coded programs and loading the desired number at the time the computer requested the information. The loaded information was redisplayed for confirmation by the Astronaut prior to being acted upon by the computer. The astronaut confirmed that he indeed wanted the displayed program to be executed by depressing a key on the keyboard.

The prelaunch alignment program was coded 01. It was inadvertently selected by the astronaut during the transearth coast. The problem that arose resulted mainly from the fact that the erasable portion of the computer memory was time-shared. The effect on the contents of the erasable memory of starting program 01 at that time was unknown. The problem was quickly dealt with by the crew and the contents of the memory verified by the ground to be correct.

The problems involving the GN&C system in the Apollo flights were minor. They provided an object lesson of the types of problems to be expected in a large program with many opportunities for error in design and operation.

The operational problems can be categorized to indicate where the operational system is most susceptible to error. The types of problems to date have been the following:

a. Ground Errors--The problems that can be categorized as ground errors included only those that arose in real time. These types of errors can be dealt with by real-time troubleshooting. Some examples have been already described.

b. Operator Errors--Although the interface between the astronauts and the guidance system had been carefully engineered during manned missions, the deficiencies of the system showed up very clearly in the examples described. The selected major real-time problems, categorized below as operator errors, clearly reflect difficulties in the design of the interactive computer programs and their use under mission conditions. These types of problems also can be easily diagnosed and corrected.

c. Misinterpretation of Design Data--The attitude and thrust-vector control systems incorporated in the Apollo guidance computer memory depend on accurate models of the spacecraft. Problems arise when the spacecraft responds to commands different from those that the computer program expects. The result can be a performance degradation resulting from either a logical error or incorrect information in the computer; both have occurred:

AS202 L/D Problem--The otherwise successful suborbital mission missed the target by 200 miles. The major cause was the lift-to-drag ratio, L/D, of of an expected 0.35 versus an actual 0.25 with the result that the vehicle had insufficient lift to attain the targeted range.

APOLLO 6 DPS Engine Shutdown--The control program for guidance during LM descent propulsion system engine operation monitored the thrust build-up after the engine had been commanded to fire. If the thrust build-up had not occurred, the program was designed to turn off the engine and generate an alarm. During the flight, the engine thrust build-up, for the first descent engine burn, did not occur at the rate expected by the program, and the computer turned off the engine. The program was designed so that appropriate real-time commands could have restarted the control program but, because of ground tracking considerations, the mission was flown with backup procedures.

d. Misinterpretation of Flight Telemetry Data--The spacecraft tele-metry data are processed by a computer complex at Houston to provide real-time displays for the flight controllers. The limitations of that system require that some data not be displayed. The display, therefore, does not give an exact picture of the spacecraft status. The prime example of how the selected displays caused mis-interpretation occurred on Apollo 8.

The flight plan of Apollo 8 called for the power to the GN & C optical subsystem to be left ON throughout the mission. The telemetry for the state of that power was not selected for real-time display. The computer monitored the sextant articulating line-of-sight angles, and this information was transmitted as part of the computer down-telemetry. Several times during the mission, the computer data indicated that the "trunnion" angle, one of the two data encoded optics system angles, changed from the expected $0^{\circ}$ to an unexpected $45^{\circ}$. This change was unexplainable from the available data. The system operation, however, indicated that by recycl-ing normal optics operating procedures the system was not affected. The decision to continue to the moon was based on that fact. Several failure models were invented during the mission to explain the problem. Later, during astronaut debriefing, it became apparent that the problem was attributed to the switching OFF of optics power. With power removed, the change in angle was to be expected each time the power was reapplied. A search through the data, which was not processed in real time, confirmed that explanation.

e. New Phenomena--To date, there have been very few surprises in the flight operations of the Apollo GN & C system. The following observations will have an effect on future GN & C design:

Visibility--Navigation in cislunar space and alignment of the iner-tial platform depend on the astronaut's identification of navigation stars. The debris generated by the spacecraft can appear in the optics as stars to make true star identification difficult. The Apollo missions, therefore, have made extensive use of the computer-inertial measurement unit combination to direct the optical line-of-sight to aid star identification.

Perigee Torquing--The size of the Apollo spacecraft resulted in considerable attitude changes in earth orbit as a result of atmospheric drag at peri-gee. This could be costly in fuel for large space stations.

f. Hardware Problems--There were very few G & N related hardware problems in the Apollo missions. The careful ground test and review of test results were the main reasons for the inflight success. Hardware problems that occurred in flight resulted in use of backup systems.

The major problem that involved the G & N was the Apollo 6 ground update procedure. The unmanned Apollo missions were dependent on ground tracking navigation data to a much greater extent than the manned missions. Several navigation updates were planned for Apollo 6. The navigation data or other remote commands to the computer were transmitted in a triple-redundant code, $K\overline{K}K$. The computer would not accept data that did not conform to this code.

During the Apollo 6 mission, several attempts to send navigation updates were rejected by the computer. The most likely cause for rejecting the data was electromagnetic interference. Review of the interface did indicate a possible problem resulting from ground command lines left disconnected at launch and unterminated. These wires were acting as the probable antenna for picking up the noise.

The source of the interference was later determined to be an ion pump associated with the fuel cells. The ion pump in the Apollo 7 spacecraft generated the same problem during a ground test in the altitude chamber. The Apollo 6 ion pump had not been ground tested in the altitude chamber. Wiring changes were also made in subsequent spacecraft to eliminate the possible noise pickup in the ground command lines.

5.4    Conclusions

(1)    Flight performance indicated that the system error mode contained in the specification was a good representation of the actual system errors during a mission. There was excellent agreement between the ground and the free-fall inertial parameter measurements.

(2)    The quality and reliability was designed and built into the equipment. With a well planned and well designed prelaunch checkout, inflight hardware problems were minimized.

(3)    Operational experience showed that automatic prelaunch checkout of space guidance, navigation, and control systems proved to be the best approach.

(4)     The mission techniques were designed after the hardware was built. Therefore, the hardware had to be flexible to accommodate different mission applications.

(5)     The complexity of the GN & C system, as well as of the total spacecraft, dictated that emphasis be placed on simulation for verification and training.

(6)     There was reasonable time available for inflight problem diagnosis, and there existed an ability for troubleshooting and diagnosis both in flight and on the ground.

(7)     Any problem found must be related (by the use of strict build control) to all possible systems, and the effects evaluated based upon requirements.

(8)     The concept of no unexplained failure was additional assurance of achieving success during the complex Apollo effort.

Section VI

CONCLUSIONS

The Guidance and Navigation (G & N) design and manufacturing effort was managed by the Guidance and Control Division of the NASA Manned Spacecraft Center, and MIT was assigned prime responsibility for all phases of the G & N design. MIT's responsibility included determination of the G & N configuration, design concepts, performance manufacturing methods and techniques, reliability and quality assurance, operation performance, computer programming, crew procedures, ground support equipment, configuration control, negotiations with the spacecraft contractors for G & N integration with the spacecraft, and technical direction to the G & N manufacturing contractors.

Soon after the design phase was initiated, MSC let out contracts to participating contractors for the manufacture of the G & N under MIT supervision. Raytheon Corporation was selected to build the computer group, Kollsman Instrument Corporation the optical units, and AC Electronics Division (formerly called AC Spark Plug Division of General Motors Corporation) was selected to build the remainder of the G & N system including the inertial subsystem, Displays and Controls, PSA, CDU, interconnecting harnesses, signal conditioning equipment, navigation base, eyepiece, and storage unit. AC Electronics Corporation was also given the task to assemble and test the complete G & N System. Separate contracts were awarded to AC Electronics Corporation and Sperry Gyroscope Corporation for the manufacture of the IRIGs and PIPAs.

Engineering personnel from these participating contractors were brought in as temporary resident engineers to assist in the design. This gave the manufacturing organizations the opportunity to become intimately familiar with the design as it developed, and to develop an awareness of the philosophies, design concepts, trade-offs and compromises that are an integral part of any development effort. The contractors in general sent well qualified senior personnel to assist in the task.

The successful design of the Apollo G & N system was dramatically demonstrated by the spectacular G & N performance evidenced throughout the Apollo flights.

Most design changes evolved from schedule constraints that demanded early releases of design to production. The schedule did not permit the luxury of testing breadboards and prototype models in the expected environments prior to design release. Much of the testing occurred after production was too far along to make indicated design changes without prohibitive expense. For this reason, changes were not made unless it could be demonstrated that the existing design simply would not suffice for the intended task. Designs were released for production before interface constraints were formalized with the spacecraft contractors, a situation that made Interface Control Document (ICD) negotiations troublesome, resulting in additional design changes and compromises, and in some instances, leading to less than optimum spacecraft configuration.

NASA, evoking the Apollo Configuration Management Manual NPC 500-1, imposed strict configuration control requirements on all Apollo contractors. Implementation of these requirements required a major documentation effort. Early designs were released with class B documentation and were not subject to formal configuration controls but were used mainly for information, system integration, and production planning. Design changes were readily made and documented as simple revisions to the drawings and specifications. With the release to manufacturing, the documentation was formalized as class A, and subject to all rigid configuration control and configuration identification restrictions.

Correspondingly, every change resulting in a piece part that was not throughly interchangeable in forward or backward production required a new part number, and additionally, part numbers were required for all related higher assemblies. This procedure led to a major documentation effort on the part of the design groups who maintained their own drawings. Moreover, complaints were registered that the design effort was being diluted by documentation.

The many changes that were made was a consequence of designs being released prior to adequate testing evaluation. The complaints notwithstanding, configuration documentation difficulties did have a stabilizing influence in the design areas. In any case, changes that were not mandatory simply were not made unless a significant benefit in cost, schedule and reliability was evident. The ultimate configuration control was effected through formal design review and change control boards.

As previously indicated, interface agreements with the spacecraft contractors were troublesome and at times difficult. NASA had labored with the contractors to establish formal interface negotiating procedures and had assisted in defining the interface areas. Preliminary information was exchanged and preliminary interface control documents were prepared. NASA then left it to the contractors to work out interface details and to document the agreed upon interfaces in approved ICDs. Dialogue was maintained between the contractor and MIT, and design information was being exchanged, but no real progress in defining the interfaces was being made at that time. Each G & N and spacecraft contractor had a constricted schedule to meet and was not willing to invalidate that schedule to accommodate another associate's design.

During the spring of 1965, very few ICDs were approved. It was becoming increasingly evident that a number of interfaces were incompatible. In the summer of 1965, NASA assembled a task force to resolve all of the interface differences. Within the span of a few weeks, all interfaces were negotiated and ICDs signed. Where design changes were indicated, implementation directives were received from NASA. Where design data were insufficient, supplements were issued.

Henceforth, NASA closely scrutinized all ICDs. Changes, as required, were promptly negotiated, thus keeping cost and schedule impacts to a minimum. The lesson to be learned from the foregoing experience is that contractors, left to their own devices, will invariably procrastinate and belabor details in defense of their own designs. What is needed is an objective arbiter to provide a continuing impetus to the program by resolving all design contentions.

The ICDs that caused the most trouble for the ISS were the prime dc power ICDs, the thermal ICDs and the ICDs that define the mechanical attachment of the G & N equipment to the vehicles.

The G & N electronics, which was designed to operate from 28 Volt $\pm$ 5% spacecraft-supplied power source, manifested problems when some circuits operated above the prescribed level and were thermally overstressed. Below that level, the circuits would be operating outside of their dynamic range, and performance would be accordingly degraded. After the design release to the manufacturer, both spacecraft contractors revealed that they could not meet the 28 V $\pm$ 5%. The problem was finally resolved by NASA when the spacecraft power systems were modified to maintain the voltage to 28.5 V $\pm$ 14%. Some G & N circuitry was also redesigned to accommodate these levels. Problems with the thermal ICDs were caused by the fact that thermal conductance between the electronic packages was less than originally designed for. This predicament necessitated relocating some of the modules in the command module PSA to reduce local hot spots. Module relocation, in turn, affected the PSA bolt mounting pattern, and as a result, the spacecraft coldplate required redesigning.

In summation, the Apollo experience has shown that MIT design capabilities in collaboration with the participating contractor manufacturing function, were a most effective means of developing the Guidance and Navigation System. Unhampered by the myriad details associated with production tooling and logistics, MIT was able to concentrate its resources of talent to produce a relatively small, effective design team. As such, MIT, unlike larger organizations, was not plagued by internal communication problems and design control bureaucracies. Moreover, because of its unique nature, MIT was in a position to monitor the activities of the participating contractors without being influenced by profit motives, thus virtually insuring a quality end item.

DISTRIBUTION LIST

Internal:

J. Alekshun

R. Booth

R. Cooper

J. Carrigan

W. Denhard

A. Freeman

J. Gilmore

E. Hall

R. Harlow

D. Hoag

F. Houston

P. Jopling

A. Laats

L. Larson

J. Lebo

R. Lones

R. McKern

G. Ogletree

R. Ragan

N. Sears

W. Stameris

R. Weatherbee

R. Woodbury

Apollo Library (40)

Tech Doc Center  (10)

External:

NASA Electronics Research Center          1 Repro + 50 Copies
Cambridge, Mass.

Attn:  M. L. Bystock

www.ingramcontent.com/pod-product-compliance
Lightning Source LLC
Chambersburg PA
CBHW051117200326
41518CB00016B/2535